Tree, Turf and Ornamental Pesticide Guide

by

W. T. THOMSON

1987 - 88 Revision

THOMSON PUBLICATIONS

P.O. BOX 9335
FRESNO, CA 93791

THIS BOOK FOR REFERENCE ONLY

READ THE LABEL CAREFULLY

PREFACE

This book is designed to be used as an everyday tool for the man making pesticide recommendations. It can be used as a fast efficient guide providing useful information as to which pesticide should be used to control a certain pest on a specific ornamental, tree or shrub. Information in this book has been obtained from the basic manufacturers labels and the EPA pesticide summaries. There will, I am sure, be various state recommendations that are not listed since they did not appear on the manufacturer's label. The book is designed only as a guide and should be counter checked by labels and local and state authorities. Read the label carefully.

This book will prove valuable to the nurseryman, landscaper, pest control operator, commercial flower grower, greenhouse owner, greenskeeper, commercial gardener and all others that are involved in the growing and maintaining of ornamentals, trees and turf. Revisions will be made periodically as deemed necessary by the author when sufficient changes have taken place in the addition and deletion of the available agricultural chemicals for use in this field.

HOW TO USE THIS BOOK

This book is designed to be as helpful as possible to the reader. After using the book the author would appreciate any of your comments on making it more useful in the future revisions.

First the book is divided into four major pesticide sections: Insecticides, Herbicides, Fungicides and Growth Regulators. The pesticides in each section are cross-referred to the plant they may be used on. If a certain pesticide is registered generally on all ornamentals it will be only listed under ornamentals, trees, shrubs, etc., not by each specific species. After knowing what pesticide may be used on a specific plant turn to the back of the particular pesticide section where the pesticides are listed alphabetically showing the particular pests each one will control. By so doing you can get a general recommendation on what material to use in each individual situation.

At the end of the chemical section is listed the address of the basic manufacturers as well as conversion and calibration charts. Remember pesticides are made to help the environment, not hurt it, so read the label carefully.

W. T. Thomson

TABLE OF CONTENTS

Page 1

Page 49

Page 93

Page 125

Page 139

INSECTICIDES

INSECTICIDES

1. **ACACIA**
   Dylox
   Meta-Systox-R
   Pentac

2. **AGAPANTHUS**
   Knox-Out

3. **AGAVE**
   (Century Plant-
   American Aloe)
   Dasanit
   Diazinon
   Dylox
   Guthion
   Malathion
   Meta-Systox-R
   Thiodan

4. **AGERATUM**
   (Floss Flower)
   Di-Syston
   Enstar
   Ficam
   Knox-Out
   Pounce
   SBP-1382
   Talstar
   Turcam

5. **AJUGA**
   Dylox
   Meta-Systox-R
   Nemacur

6. **ALDER**
   Dylox
   Meta-Systox-R
   Orthene

7. **ALMOND**
   (Flowering)
   Ficam-W

   Trithion
   Turcam

8. **ALOE**
   Dasanit
   Diazinon
   Dylox
   Guthion
   Malathion
   Meta-Systox-R
   Thiodan

9. **ALUMINUM
   PLANT**
   (Pilea)
   Enstar
   Knox-Out
   Methomyl
   Oxamyl

10. **ALYSSUM**
    Diazinon
    Dylox
    Guthion
    Knox-Out
    Malathion
    Meta-Systox-R
    Pounce
    Talstar
    Thiodan

11. **AMARYLLIS**
    Trithion

12. **ANEMONE**
    Pounce

13. **ANTHURIUM**
    Delnav
    Diazion
    Dylox
    Guthion
    Malathion

    Meta-Systox-R
    Nemacur
    Orthene
    Thiodan

14. **APHELANDRA**
    (Zebra Plant)
    Enstar
    Pentac
    Pounce
    Vydate

15. **APISCIA**
    Delnav

16. **APPLE**
    (Flowering)
    Azodrin
    Diazinon
    Dibrom
    Dylox
    Guthion
    Malathion
    Meta-Systox-R
    Oxamyl
    Thiodan
    Volck Oil
    Vydate

17. **ARALIA**
    Delnav
    Diazinon
    Dibrom
    Dylox
    Enstar
    Guthion
    Knox-Out
    Malathion
    Meta-Systox R
    Thiodan

18. **ARBOR VITAE**
    (Thuja)
    Azodrin
    Baytex (Entex)
    Cygon (Dimethoate)
    Diazion
    Dibrom
    Dylox
    Ethion
    Guthion
    Imidan
    Lindane
    Malathion
    Meta-Systox-R
    Methomyl
    Morestan
    Orthene
    Pentac
    Sevin
    Talstar
    Thiodan
    Trithion
    Zolone

19. **ARDISIA**
    Dasanit
    Diazinon
    Dylox
    Guthion
    Malathion
    Meta-Systox-R
    Methomyl
    Oxamyl
    Pounce
    Thiodan

20. **ARROWHEAD**
    Knox-Out

21. **ASARINA**
    Ficam-W
    Turcam

22. **ASH**
    Baytex (Entex)
    Diazinon

Dibrom
Dursban
Dylox
Ethion
Guthion
Imidan
Malathion
Meta-Systox-R
Orthene
Thiodan
Trithon

23. **ASPEN**
    Baytex (Entex)
    Diazinon
    Dylox
    Guthion
    Malathion
    Meta-Systox-R
    Sevin
    Thiodan

24. **ASTER**
    Diazinon
    Di-Syston
    Dylox
    Guthion
    Lindane
    Malathion
    Meta-Systox-R
    Nicotine
    Orthene
    Pyrenone
    SBP-1382
    Talstar
    Trithion

25. **ASTRID**
    Oxamyl

26. **AUCUBA**
    (Gold Dust Plant)
    Dibrom
    Meta-Systox-R
    Pentac
    Vydate

27. **AZALEA**
    Azodrin
    Cygon (Dimethoate)
    Dasanit
    Diazinon
    Dibrom
    Di-Syston
    Dylox
    Enstar
    Ethion
    Ficam-W
    Guthion
    Knox-Out
    Lindane
    Malathion
    Meta-Systox-R
    Methyl Bromide
    Mocap
    Nemacur
    Oil
    Orthene
    Oxamyl
    Pentac
    Pounce
    Pyrenone
    Sevin
    SBP-1382
    Talstar
    Thimet
    Thiodan
    Trithion
    Turcam
    Vydate

28. **BARBERRY**
    Dasanit
    Diazinon
    Dylox
    Guthion
    Malathion
    Meta-Systox-R
    Methomyl
    Oxamyl
    Talstar
    Thiodan

29. **BEAUTY BERRY**
Methomyl

30. **BEECH**
Diazinon
Dylox
Guthion
Imidan
Lindane
Malathion
Meta-Systox-R
Orthene
Thiodan
Trithion

31. **BEGONIA**
Baytex (Entex)
Diazinon
Dylox
Endrin
Ficam-W
Guthion
Knox-Out
Malathion
Meta-Systox-R
Nicotine
Pounce
Pyrenone
SBP-1382
Talstar
Thiodan
Trithion
Turcam

32. **BIRCH**
Azodrin
Baytex (Entex)
Cygon (Dimethoate)
Diazinon
Dibrom
Di-Syston
Dylox
Ethion
Guthion
Imidan
Lindane

Malathion
Meta-Systox-R
Orthene
Pentac
Sevin
Temik
Thimet
Thiodan
Volck Oil

33. **BIRD OF PARADISE**
Trithion

34. **BISHOPWOOD TREE**
Methomyl

35. **BLUE MIST**
Dasanit
Diazinon
Dylox
Guthion
Malathion
Meta-Systox-R
Thiodan

36. **BOTTLEBRUSH**
Knox-Out

37. **BOUGANVILLAE**
Dasanit
Diazinon
Dylox
Ficam-W
Guthion
Malathion
Meta-Systox-R
Methomyl
Thiodan

38. **BOX ELDER (Acre)**
Diazinon
Dibrom
Dylox

Guthion
Malathion
Meta-Systox-R
Orthene
Thiodan

39. **BOXWOOD (Buxus)**
Azodrin
Cygon (Dimethoate)
Dasanit
Diazinon
Dibrom
Dylox
Ethion
Ficam-W
Guthion
Lindane
Malathion
Meta-Systox-R
Methomyl
Mocap
Nemacur
Nicotine
Oxamyl
Talstar
Thimet
Thiodan
Trithion
Turcam
Vydate

40. **BROMELIADS**
Enstar
Mocap
Pentac

41. **BROOM**
Oxamyl

42. **BUCKEYE**
Methomyl

43. **BUILDINGS & GREENHOUSES (Inside & Outside)**
Altosid
Baygon
Baytex
Borax
Boric Acid
Bromyl
Chlordane
Chloropicrin
Cygon
Cythion
Delnav (Deltic)
Diazinon
Dibrom
Dri-Die (Silica-Gel)
Dursban
Dylox
Enstar
Ethion
Ficam-W
Guthion
Kelthane
Lindane
Malathion
Mesurol
Meta-Systox-R
Methoxychlor
Methyl Bromide
Nicotine
Pentac
Pyrethrins
Sevin
Sodium Floride
Sodium Fluosilicate
SBP-1382
Thiodan
Vapona
Vikane

44. **BUTTERFLY BUSH**
Methomyl

45. **CACTUS**
Knox-Out
Mocap
Nemacur
Orthene
Talstar
Trithion
Vydate

46. **CALADIUM**
Dasanit
Diazinon
Dylox
Guthion
Malathion
Meta-Systox-R
Mocap
Oxamyl
Talstar
Thiodan

47. **CALENDULA**
Di-Syston
Ficam-W
Knox-Out
Meta-Systox-R
Orthene
Pounce
Rotenone
SBP-1382
Talstar
Turcam

48. **CALYCANTHUS**
Methomyl
Oxamyl

49. **CAMELLIA**
Azodrin
Cygon (Dimethoate)
Dasanit
Diazinon
Dibrom
Di-Syston
Dylox
Ethion

Guthion
Knox-Out
Lindane
Malathion
Meta-Systox-R
Mocap
Pyrenone
Thiodan
Trithion

50. **CANDY TUFT**
Diazinon
Dylox
Guthion
Malathion
Meta-Systox-R
Thiodan

51. **CANNA**
Nicotine
Rotenone

52. **CARNATIONS (Dianthus)**
Azodrin
Cygon (Dimethoate)
Delnav
Diazinon
Dibrom
Di-Syston
Dylox
Enstar
Ficam-W
Guthion
Kelthane
Knox-Out
Malathion
Meta-Systox-R
Morocide
Nicotine
Omite
Orthene
Oxamyl
Pentac
Phosdrin
Plictran

Pounce
Pyrethrins
Sevin
Talstar
Tedion
Temik
Thimet
Thiodan
Trithion
Turcam
Vapona
Vydate

53. **CATALPA**
Diazinon
Dylox
Ethion
Guthion
Malathion
Meta-Systox-R
Thiodan

54. **CEDAR**
Cygon (Dimethoate)
Diazinon
Dibrom
Dylox
Ethion
Ficam-W
Guthion
Imidan
Malathion
Meta-Systox-R
Methomyl
Nicotine
Orthene
Pentac
Talstar
Thiodan
Trithion
Turcam

55. **CELOSIA**
(Cockcombs)
Ficam-W
Knox-Out

Talstar
Turcam

56. **CHENILE PLANT**
Talstar

57. **CHERRY**
(Flowering)
Dasanit
Diazinon
Dursban
Dylox
Guthion
Malathion
Meta-Systox-R
Nemacur
Orthene
Oxamyl
Thiodan
Trithion
Volck Oil
Vydate

58. **CHESTNUT**
Diazinon
Dylox
Guthion
Malathion
Meta-Systox-R
Sulfur
Thiodan

59. **CHINESE**
EVERGREEN
(Aglaonema)
Delnav
Diazinon
Dylox
Guthion
Malathion
Meta-Systox-R
Mocap
Pentac
Thiodan

60. **CHRISTMAS**
BERRY
Dasanit
Diazinon
Dylox
Guthion
Malathion
Meta-Systox-R
Thiodan

61. **CHRYSANTHE-**
MUMS
Azodrin
Bacillus
Thuringiensis
Baytex (Entex)
Dasanit
Delnav
Diazinon
Dimilan
Di-Syston
Dursban
Dylox
Enstar
Ficam-W
Guthion
Kelthane
Knox-Out
Malathion
Meta-Systox-R
Methomyl
Methoxychlor
Minex
Morocide
Nicotine
Omite
Orthene
Oxamyl
Parathion
Pentac
Permethrin
Phosdrin
Pirimor
Plictran
Pounce
Pyrethrins

SBP-1382
Sevin
Systox
Talstar
Tedion
Temik
Thimet
Thiodan
Trithion
Turcam
Vydate

62. **CINERARIA**
Pounce
Talstar
Trithion

63. **CISSUS**
(Grape Ivy)
Pounce

64. **CLEMATIS**
Nemacur

65. **COLEUS**
Diazinon
Dylox
Ficam-W
Guthion
Knox-Out
Malathion
Meta-Systox-R
Methomyl
Orthene
Pounce
SBP-1382
Talstar
Thiodan
Trithion
Turcam

66. **COLUMBINE**
Nicotine
Orthene

67. **CORDYLINE**
Talstar

68. **CORYLUSAVELL**
ANA
Talstar

69. **COSMOS**
Diazinon
Dylox
Guthion
Malathion
Meta-Systox-R
Nicotine
Thiodan
Trithion

70. **COTONEASTER**
Ethion
Knox-Out
Methomyl
Nemacur
Talstar
Trithion

71. **COTTONWOOD**
Diazinon
Dursban
Dylox
Furadan
Guthion
Malathion
Meta-Systox-R
Talstar
Thimet
Thiodan

72. **CRABAPPLE**
Dibrom
Dylox
Ficam-W
Meta-Systox-R
Nemacur
Oil
Orthene
Turcam

73. **CREEPING FIG**
Pounce

74. **CRAPE MYRTLE**
Bidrin
Knox-Out
Methomyl

75. **CROSSANDRA**
Talstar

76. **CROTONS**
Dasanit
Delnav
Diazinon
Dylox
Enstar
Guthion
Knox-Out
Malathion
Meta-Systox-R
Methomyl
Oxamyl
Pentac
Pounce
Thiodan
Trithion
Vydate

77. **CYCLAMEN**
Dasanit
Diazinon
Dylox
Endrin
Guthion
Malathion
Meta-Systox-R
Talstar
Thiodan

78. **CYPERUS**
(Umbrella plant)
Cygon (Dimethoate)
Diazinon
Dylox
Guthion

8

Malathion
Meta-Systox-R
Oxamyl
Thiodan

79. CYPRESS
Dasanit
Diazinon
Dylox
Guthion
Malathion
Meta-Systox-R
Methomyl
Orthene
Pentac
Talstar
Thiodan

80. DAFFODIL
Diazinon
Dylox
Ficam-W
Guthion
Lindane
Malathion
Systox
Thiodan
Turcam

81. DAHLIA
Delnav
Diazinon
Dibrom
Di-Syston
Dylox
Ficam-W
Guthion
Lindane
Malathion
Meta-Systox-R
Methoxychlor
Nicotine
Pyrethrins
Talstar
Temik
Thimet

Thiodan
Trithion
Turcam

82. DAISY
Azodrin
Diazinon
Dylox
Ficam-W
Guthion
Knox-Out
Malathion
Meta-Systox-R
Orthene
Thiodan
Turcam

83. DELPHINIUM (Larkspur)
Diazinon
Di-Syston
Dylox
Guthion
Lindane
Malathion
Meta-Systox-R
Nicotine
Pentac
Thiodan
Trithion

84. DEUTZIA
Methomyl

85. DEVILWOOD
Oxamyl

86. DICHLORI-SANDRAS
Delnav
Diazinon
Dylox
Guthion
Malathion
Meta-Systox-R
Thiodan

87. DIEFFENBACHIA
Enstar
Methomyl
Orthene
Oxamyl
Pentac
Talstar
Vydate

88. DOGWOOD
Baytex (Entex)
Diazinon
Dibrom
Dursban
Dylox
Ethion
Guthion
Imidan
Knox-Out
Lindane
Malathion
Meta-Systox-R
Pentac
Pyrethrins
Sevin
Talstar
Thiodan
Volck Oil

89. DRACAENA
Enstar
Methomyl
Oxamyl
Pentac
Talstar
Vydate

90. DUSTY MILLER
Pentac
Talstar

91. ELM
Diazinon
Dibrom
Di-Syston

9

Dursban
Dylox
Ethion
Furadan
Guthion
Imidan
Lindane
Malathion
Meta-Systox-R
Oil
Pentac
Thiodan
Trithion

92. **ESCALVONIA**
Knox-Out

93. **EUCALYPTUS-GUM**
Baytex (Entex)
Diazinon
Dylox
Guthion
Malathion
Meta-Systox-R
Thiodan
Trithion

94. **EUONYMUS**
Azodrin
Baytex (Entex)
Cygon
Dasanit
Diazinon
Dibrom
Dimethoate
Di-Syston
Dylox
Ethion
Ficam-W
Guthion
Knox-Out
Malathion
Meta-Systox-R
Methomyl

Nemacur
Orthene
Oxamyl
Talstar
Thiodan
Trithion
Turcam
Vydate

95. **EURYA**
Dasanit
Diazinon
Dylox
Guthion
Malathion
Meta-Systox-R
Thiodan

96. **FEIJOA-PINEAPPLE GUAVA**
Diazinon
Dylox
Guthion
Malathion
Meta-Systox-R
Thiodan

97. **FERNS**
Delnav
Dylox
Enstar
Ficam-W
Guthion
Knox-Out
Malathion
Meta-Systox-R
Methomyl
Mocap
Nemacur
Nicotine
Oxamyl
Pentac
Pounce
Talstar

Thiodan
Trithion
Turcam
Vydate

98. **FICUS-RUBBER PLANT**
Cygon
Diazinon
Dimethoate
Dylox
Enstar
Ethion
Ficam-W
Guthion
Malathion
Meta-Systox-R
Orthene
Pentac
Pyrethrins
Talstar
Thiodan
Trithion
Turcam

99. **FIR**
Azodrin
Cygon
Diazinon
Dibrom
Dimethoate
Dylox
Ethion
Guthion
Imidan
Malathion
Meta-Systox-R
Methomyl
Oxamyl
Pentac
Pounce
Sevin
Talstar
Thiodan
Trithion

100. **FITTONIA**
Methomyl
Pounce

101. **FOLIAGE
PLANTS
(All)**
Azodrin
Orthene
Zectran

102. **FORSYTHIA**
Dylox
Meta-Systox-R

103. **FRIENDSHIP
PLANT**
Knox-Out

104. **FUCHSIA**
Azodrin
Dylox
Enstar
Ficam-W
Meta-Systox-R
Pounce
SBP-1382
Trithion
Turcam

105. **GALLBERRY**
Dasanit
Diazinon
Dylox
Guthion
Malathion
Meta-Systox-R
Thiodan

106. **GARDENIA**
Azodrin
Cygon
Dasanit
Dibrom
Dimethoate
Dylox

Ethion
Guthion
Knox-Out
Lindane
Malathion
Meta-Systox-R
Mocap
Nemacur
Oil
Orthene
Oxamyl
Pentac
Talstar
Thiodan
Trithion
Vydate

107. **GAZANIA**
Ficam-W
Knox-Out
Pounce
Turcam

108. **GERANIUM**
Azodrin
Baytex (Entex)
Diazinon
Dylox
Enstar
Ficam-W
Guthion
Knox-Out
Malathion
Meta-Systox-R
Nicotine
Oxamyl
Pyrethrins
Rotenone
SBP-1382
Talstar
Thiodan
Turcam

109. **GERBERA-
TRANSVAAL
DAISY**
Cygon
Diazinon
Dimethoate
Dylox
Enstar
Guthion
Malathion
Meta-Systox-R
Temik
Thiodan

110. **GIRIKO**
Methomyl
Oxamyl

111. **GLADIOLUS**
Cygon
Dasanit
Delnav
Diazinon
Dimethoate
Di-Syston
Dylox
Guthion
Lindane
Malathion
Meta-Systox-R
Methomyl
Nicotine
Orthene
Oxamyl
Parathion
Pounce
Pyrenone
Sevin
Thiodan
Trithion
Vydate

112. **GLOXINIA**
Azodrin
Dasanit
Diazinon

Dylox
Enstar
Guthion
Malathion
Meta-Systox-R
Talstar
Thiodan

113. **GOLD DUST TREE**
Knox-Out
Oxamyl

114. **GOLD FISH PLANT**
Oxamyl

115. **GUAVA**
Knox-Out
Trithion

116. **GYPSOPHILA (Baby's Breath)**
Ficam-W
Orthene
Oxamyl
Pounce
Vydate

117. **HAWTHORN**
Diazinon
Dylox
Guthion
Imidan
Malathion
Meta-Systox-R
Orthene
Talstar
Thiodan
Trithion
Volck Oil

118. **HEART LEAF**
Oxamyl

119. **HEATHER**
Dylox
Meta-Systox-R
Trithion

120. **HEDERA**
Talstar

121. **HEMLOCK**
Acaraben
Cygon
Diazinon
Dibrom
Dimethoate
Dylox
Ethion
Guthion
Imidan
Malathion
Meta-Systox-R
Methomyl
Oil
Orthene
Pentac
Sevin
Talstar
Thiodan
Trithion

122. **HICKORY**
Imidan
Orthene

123. **HIBISCUS-ROSE MALLOW**
Dasanit
Dibrom
Dylox
Guthion
Lindane
Malathion
Meta-Systox-R
Nemacur
Orthene
Oxamyl
Pentac

Talstar
Thiodan

124. **HINDU-ROPE**
Vydate

125. **HOLLY (Ilex)**
Cygon
Dasanit
Diazinon
Dibrom
Di-Syston
Dylox
Ethion
Ficam-W
Guthion
Knox-Out
Malathion
Meta-Systox-R
Methomyl
Mocap
Nemacur
Oil
Orthene
Oxamyl
Talstar
Temik
Thimet
Thiodan
Trithion
Turcam
Vydate

126. **HOLLYHOCKS (Althaea)**
Nicotine
Trithion

127. **HONEYSUCKLE**
Dasanit
Diazinon
Dylox
Guthion
Knox-Out
Malathion
Thiodan

128. **HYPOESTES**
(Baby-Tears)
Ficam-W
Turcam

129. **HOYA**
(Wax Plant)
Knox-Out
Methomyl
Oxamyl
Vydate

130. **HYACINTH**
Diazinon
Guthion
Malathion
Thiodan

131. **HYDRANGEA**
Diazinon
Dylox
Enstar
Ficam-W
Guthion
Knox-Out
Malathion
Pentac
Sevin
Thiodan
Trithion
Turcam

132. **HYPERICUM**
Pounce

133. **ICE PLANT**
Orthene

134. **IMPATIENS**
Ficam-W
Knox-Out
Talstar
Turcam

135. **IRIS**
Cygon

Dasanit
Diazinon
Dimethoate
Di-Syston
Dylox
Ficam-W
Guthion
Malathion
Nemacur
Nicotine
Oxamyl
Thiodan
Turcam

136. **IVY**
Diazinon
Dylox
Endrin
Enstar
Ficam-W
Guthion
Knox-Out
Malathion
Meta-Systox-R
Methomyl
Nemacur
Orthene
Oxamyl
Pentac
Pounce
SBP-1382
Talstar
Thiodan
Trithion
Turcam

137. **IXORA**
Diazinon
Dylox
Ethion
Guthion
Malathion
Meta-Systox-R
Methomyl
Pentac
Thiodan

138. **JADE PLANT**
(Crassula)
Knox-Out
Oxamyl
Pounce

139. **JAPANESE**
PLUM
Dasanit

140. **JASMINE**
Dasanit
Diazinon
Dylox
Guthion
Knox-Out
Malathion
Meta-Systox-R
Methomyl
Pentac
Thiodan
Trithion

141. **JUNGLE**
FLAME
Dasanit
Diazinon
Dylox
Guthion
Malathion
Meta-Systox-R
Thiodan

142. **JUNIPERS**
Azodrin
Baytex (Entex)
Cygon
Dasanit
Delnav
Diazinon
Dibrom
Dimethoate
Dylox
Ethion
Ficam-W
Guthion

Imidan
Knox-Out
Lindane
Malathion
Meta-Systox-R
Methomyl
Morestan
Nemacur
Oxamyl
Pentac
Sevin
Talstar
Thiodan
Trithion
Turcam
Vydate

143. **KALANCHOE**
Enstar
Talstar

144. **KALANDRA**
Ficam-W
Turcam

145. **LAGUSTRUM**
Talstar

146. **LANTANA**
Enstar
Talstar
Trithion

147. **LAUREL-CHERRY**
Dasanit

148. **LAUREL-SWEET BAY**
Diazinon
Dylox
Guthion
Malathion
Meta-Systox-R
Methomyl

Pounce
Talstar
Thiodan

149. **LEUCOTHOE**
Knox-Out
Methomyl

150. **LILAC**
Diazinon
Di-Syston
Dursban
Dylox
Ethion
Ficam-W
Guthion
Knox-Out
Lindane
Malathion
Meta-Systox-R
Nicotine
Oil
Sevin
Thiodan
Trithion
Turcam

151. **LILIES**
Baytex (Entex)
Cygon
Dasanit
Diazinon
Dimethoate
Di-Syston
Enstar
Ficam-W
Guthion
Malathion
Meta-Systox-R
Nemacur
Nicotine
Parathion
Systox
Talstar
Temik
Thimet

Thiodan
Trithion
Turcam

152. **LINDEN BASSWOOD**
Diazinon
Dylox
Guthion
Malathion
Meta-Systox-R
Trithion

153. **LITHODORA**
Talstar

154. **LOBELIA**
Talstar

155. **LOCUST**
Diazinon
Dibrom
Di-Syston
Dylox
Guthion
Malathion
Meta-Systox-r
Orthene
Thiodan
Trithion

156. **LOQUAT**
Talstar

157. **MAGNOLIA**
Azodrin
Dasanit
Diazinon
Dibrom
Dylox
Ethion
Guthion
Malathion
Meta-Systox-R
Methomyl
Oxamyl

Pounce
Thiodan
Trithion
Volck Oil
Vydate

158. **MALUS**
Knox-Out

159. **MAPLES**
Azodrin
Baytex (Entex)
Diazinon
Dibrom
Dylox
Ethion
Guthion
Imidan
Lindane
Malathion
Meta-Systox-R
Orthene
Sevin
Talstar
Thiodane
Volck Oil

160. **MARANTA**
(Arrow Root)
Knox-Out
Oxamyl
Pentac
Vydate

161. **MARIGOLD**
Baytex (Entex)
Diazinon
Dibrom
Dylox
Enstar
Ficam-W
Guthion
Knox-Out
Malathion

Meta-Systox-R
Nicotine
Pounce
Pyrethrins
SBP-1382
Talstar
Thiodan
Turcam

162. **MEDICINE**
PLANT
Oxamyl

163. **MESEMPRY-**
ANTHEMUM
Talstar

164. **METRO-**
SIDEROS
(Bottle-Brush)
Dasanit
Diazinon
Dylox
Guthion
Malathion
Meta-Systox-R
Thiodan

165. **MIMOSA**
Ficam-W
Turcam

166. **MOCK ORANGE**
Diazinon
Dylox
Ethion
Guthion
Malathion
Meta-Systox-R
Methomyl
Oxamyl
Thiodan

167. **MONSTERA**
Talstar

168. **MOONFLOWER**
Trithion

169. **MULBERRY**
Oxamyl

170. **MYRTLE**
Ficam-W
Methomyl
Trithion
Turcam

171. **NANDINA**
Knox-Out
Talstar

172. **NARCISSUS**
Diazinon
Dylox
Guthion
Malathion
Meta-Systox-R
Nemacur
Proxol
Thiodan

173. **NASTURTIUM**
Azodrin
Diazinon
Dylox
Ficam-W
Guthion
Malathion
Meta-Systox-R
Nicotine
Thiodan
Trithion
Turcam

174. **NEPHTHYTIS**
Pounce

175. **NERVE PLANT**
Oxamyl

176. NICOTIANA
Ficam-W
Turcam

177. NURSERY STOCK
(All kinds)
Azodrin
D-D
Diazinon
Dibrom
Di-Syston
Dylox
Endrin
Ethion
Furadan
Guthion
Kelthane
Lindane
Malathion
Mavrik
Meta-Systox-R
Mocap
Nemacur
Proxol
Sevin
Systox
Tedion
Thiodan
Thuricide
Vapona
Volck Oil
Zectran

178. OAK
Baytex (Entex)
Cygon
Diazinon
Dibrom
Dimethoate
Dylox
Ethion
Guthion
Imidan
Lindane
Malathion

Meta-Systox-R
Orthene
Sevin
Talstar
Thiodan
Volck Oil

179. OLEANDER
Dylox
Ficam-W
Knox-Out
Meta-Systox-R
Trithion

180. OLIVE
Talstar

181. ORCHIDS
Azodrin
Delnav
Dylox
Guthion
Kelthane
Knox-Out
Malathion
Meta-Systox-R
Orthene
Pyrethrins
Temik
Thiodan
Trithion

182. OREGON GRAPE
(Mahonia)
Dasanit
Diazinon
Dylox
Guthion
Malathion
Meta-Systox-R
Thiodan

183. ORNAMENTALS
(All kinds)
Altocid
Avid
Azodrin
Baytex
BHC
Bacillus
Thuringiensis
Baygon
Calcium Cyanide
Chloropicirin
Cythion
D-D
Diazinon
Dibrom
Di-Syston
Dormant Oils
Dursban
Dylox
EDB
Enstar
Ethion
Guthion
Imidan
Kelthane
Kyrocide
Lindane
Malathion
Mavrik
Mesurol
Meta-Systox-R
Methaldehyde
Methomyl
Methoxychlor
Methyl Bromide
Morestan
Nicotine
Nosema Spore
Parathion
Plictran
Proxol
Pydrin
Pyrethins
Sevin
Systox

16

Talstar
Tedion
Telone
Thiodan
Vapam
Vapona
Vendex
Vorlex
Zectran

184. **OSMANTHUS**
(Holly-Olive)
Dasanit
Diazinon
Dylox
Guthion
Knox-Out
Malathion
Meta-Systox-R
Methomyl
Thiodan

185. **PACHYS-**
ANDRA
Ficam-W
Knox-Out
Nemacur
Turcam

186. **PALMS**
Enstar
Methomyl
Oxamyl
Pentac
Talstar
Trithion
Vydate

187. **PANSY**
Diazinon
Dylox
Ficam-W
Guthion
Malathion
Meta-Systox-R
Nicotine

Orthene
Pounce
Talstar
Thiodan
Trithion
Turcam

188. **PEACH**
(Flowering)
Dursban
Dylox
Oxamyl
Trithion
Vydate

189. **PELEGONIUM**
Talstar

190. **PEONIES**
Baytex (Entex)
Diazinon
Dylox
Guthion
Malathion
Meta-Systox-R
Oxamyl
Thiodan
Vydate

191. **PEPEROMIA**
Delnav
Diazinon
Dylox
Enstar
Guthion
Malathion
Meta-Systox-R
Methomyl
Oxamyl
Pentac
Thiodan
Vydate

192. **PERIWINKLE**
(Vinca)
Diazinon

Dylox
Guthion
Knox-Out
Malathion
Meta-Systox-R
Nemacur
Pounce
Talstar
Thiodan

193. **PETUNIA**
Baytex (Entex)
Diazinon
Di-Syston
Dylox
Enstar
Ficam-W
Guthion
Knox-Out
Malathion
Meta-Systox-R
Nicotine
Pounce
SBP-1382
Talstar
Thiodan
Turcam

194. **PHILODEN-**
DRON
Azodrin
Delnav
Diazinon
Dylox
Enstar
Guthion
Malathion
Meta-Systox-R
Methomyl
Mocap
Oxamyl
Pentac
Thiodan
Vydate

195. **PHLOX**
Baytex (Entex)
Diazinon
Dylox
Guthion
Malathion
Meta-Systox-R
Nicotine
Talstar
Thiodan
Trithion

196. **PHOTINIA**
Dasanit
Diazinon
Dylox
Guthion
Knox-Out
Malathion
Meta-Systox-R
Methomyl
Talstar
Thiodan

197. **PIERIS**
Methomyl
Nemacur

198. **PIGGY BACK PLANT**
Pounce

199. **PINES**
Azodrin
Cygon
Diazinon
Dibrom
Dimethoate
Di-Syston
Dursban
Dylox
Ethion
Ficam-W
Furadan
Guthion
Imidan

Lindane
Malathion
Meta-Systox-R
Methomyl
Methyl Parathion
Oil
Orthene
Oxamyl
Nemacur
Pentac
Pounce
Pydrin
Talstar
Thimet
Thiodan
Trithion
Turcam

200. **PINK POLKA DOT**
Oxamyl

201. **PITTOSPORUM**
Diazinon
Dylox
Ethion
Guthion
Knox-Out
Malathion
Meta-Systox-R
Oxamyl
Pentac
Talstar
Thiodan
Trithion

202. **PLUM (Flowering)**
Dasanit
Diazinon
Dibrom
Dursban
Dylox
Guthion
Malathion

Meta-Systox-R
Orthene
Oxamyl
Thiodan
Trithion
Vydate

203. **PODOCARPUS**
Delnav
Diazinon
Dylox
Guthion
Malathion
Meta-Systox-R
Methomyl
Pentac
Thiodan

204. **POINSETTIAS**
Cygon
Dibrom
Dimethoate
Dylox
Guthion
Malathion
Meta-Systox-R
Methomyl
Orthene
Pentac
Pounce
Plictran
SBP-1382
Talstar
Temik
Thiodan
Trithion

205. **PONDS & LAKES**
Abate
Altosid
BTI
Baygon
Bayluscide
Baytex
Copper Sulfate

18

Diazinon
Dibrom
Dursban
Dylox
EPN
Guthion
Malathion
Meta-Systox-R
Methyl Parathion
Parathion
Thiodan

206. **POPLAR**
Diazinon
Dibrom
Dylox
Ethion
Guthion
Malathion
Meta-Systox-R
Orthene
Thiodan
Trithion

207. **PORTULACA**
Ficam-W
Turcam

208. **POTENTILLA**
Talstar

209. **PREMISES**
Abate
Affirm
Allethrin
Altosid
Amdro
BTI
Baygon
Baytex
Borax
Boric Acid
Bromyl
Chlordane
Cygon
Cythion

DDVP
Delnav (Deltic)
Diazinon
Dibrom
Dimethoate
Dimilin
Dursban
Dylox
EPN
Ficam
Furadan
Guthion
Heptachlor
Kelthane
Lindane
Logic
Malathion
Mesurol
Meta-Systox-R
Methoxychlor
Nicotine
Nosema Spore
Permethrin
Pharorid
Precor
Pro-Drone
Pydrin
Pyrenone
Pyrethrins
Rotenone
SBP-1382
Safrotin
Sevin
Sulfur
Sumithion
Thiodan
Turcam
Vapona

210. **PRIMROSE**
Diazinon
Dylox
Guthion
Knox-Out
Malathion
Meta-Systox-R

Pounce
Rotenone
Systox
Talstar
Thiodan
Trithion

211. **PRIVET-**
LIGUSTRUM
Azodrin
Dasanit
Diazinon
Dibrom
Dylox
Ethion
Guthion
Knox-Out
Lindane
Malathion
Meta-Systox-R
Methomyl
Oxamyl
Talstar
Thiodan
Trithion
Volck Oil

212. **PROTEAS**
Nemacur

213. **PURPLE**
PASSION
Enstar
Methomyl

214. **PYRACANTHA**
(Firethorne)
Dasanit
Diazinon
Di-Syston
Dylox
Ficam-W
Guthion
Knox-Out
Lindane
Malathion

19

Meta-Systox-R
Nemacur
Orthene
Oxamyl
Talstar
Tedion
Thiodan
Trithion
Turcam
Volck Oil

215. RANUNCULUS
Pounce

216. RAPHIOLEPSIS
Knox-Out

217. REDWOOD
Methomyl

218. RHODODEN-DRON
Diazinon
Di-Syston
Dursban
Dylox
Ficam-W
Guthion
Knox-Out
Lindane
Malathion
Meta-Systox-R
Methomyl
Nemacur
Thiodan
Trithion
Turcam

219. ROSES
Azodrin
Baytex (Entex)
Cygon
Dasanit
Delnav
Diazinon

Dibrom
Dimethoate
Di-Syston
Dylox
Ethion
Ficam-W
Guthion
Karathane
Kelthane
Knox-Out
Lindane
Malathion
Meta-Systox-R
Methomyl
Methoxychlor
Morestan
Nemacur
Nicotine
Omite
Orthene
Oxamyl
Pentac
Phosdrin
Plictran
Pounce
Pyrethrins
Sevin
SBP-1382
Talstar
Tedion
Temik
Thiodan
Turcam
Vydate
Zolone

220. ROTUNDIFOLIA
Vydate

221. RUSSIAN OLIVE
Dasanit

222. SAGE
Knox-Out

223. SALVIA
Dylox
Ficam-W
Knox-Out
Meta-Systox-R
Pounce
SBP-1382
Talstar
Turcam

224. SANSERVERIA-BOWSTRING HEMP
Dasanit
Diazinon
Dylox
Guthion
Knox-Out
Malathion
Meta-Systox-R
Methomyl
Mocap
Oxamyl
Thiodan
Vydate

225. SASANQUA
Ethion

226. SCHEFFLERA
Ficam-W
Pentac
Talstar
Turcam

227. SCINDAPSUS
(Pothos)
Dasanit
Diazinon
Dylox
Guthion
Malathion
Meta-Systox-R
Methomyl
Pentac

Talstar
Thiodan

228. SEA GRAPE
Methomyl

229. SEDUM
Nemacur

**230. SEERSUCKER
PLANTS**
Delnav
Diazinon
Dylox
Guthion
Malathion
Meta-Systox-R
Thiodan

**231. SHRUBS
(All kinds)**
Azodrin
D-D
Diazinon
Dibrom
Di-Syston
Dursban
Dylox
Endrin
Ethion
Guthion
Kelthane
Lindane
Malathion
Mavrik
Mesurol
Meta-Systox-R
Methomyl
Methoxychlor
Methyl Parathion
Morestan
Nicotine
Orthene
Proxol
Systox
Talstar

Tedion
Thiodan
Thuricide
Trithion
Volck Oil
Zectran

232. SMOKE TREE
Talstar

233. SNAKE PLANT
Knox-Out

234. SNAPDRAGONS
Baytex (Entex)
Diazinon
Dibrom
Dylox
Endrin
Enstar
Ficam-W
Guthion
Malathion
Meta-Systox-R
Nicotine
Orthene
Oxamyl
Pentac
Pounce
Pyrethrins
Talstar
Temik
Thiodan
Trithion
Turcam
Vydate

**235. SPIDER PLANT
(Cleome)**
Knox-Out
Methomyl
Nicotine

236. SPIREA
Diazinon
Di-Syston

Dylox
Ficam-W
Guthion
Lindane
Malathion
Meta-Systox-R
Methomyl
Nicotine
Orthene
Talstar
Thiodan
Turcam

237. SPRUCE
Azodrin
Baytex (Entex)
Diazinon
Dibrom
Dylox
Ethion
Guthion
Imidan
Malathion
Meta-Systox-R
Methomyl
Oil
Orthene
Pentac
Pounce
Sevin
Talstar
Thiodan
Trithion

238. STALICE
Orthene

239. STOCKS
Diazinon
Dibrom
Dylox
Guthion
Malathion
Meta-Systox-R
Pounce
Pyrethrins

21

Thiodan
Thiram

240. **STRAW-**
   FLOWER
   Knox-Out
   Orthene

241. **SUNFLOWER**
   Diazinon
   Dylox
   Guthion
   Malathion
   Meta-Systox-R
   Thiodan

242. **SWEET FEIJON**
   Vydate

243. **SWEET GUM**
   Talstar

244. **SWEET PEA**
   Diazinon
   Dylox
   Guthion
   Kelthane
   Malathion
   Meta-Systox-R
   Nicotine
   Pyrethrins
   Thiodan
   Trithion

245. **SYCAMORE**
   Diazinon
   Dibrom
   Dylox
   Guthion
   Lindane
   Malathion
   Meta-Systox-R
   Orthene
   Thiodan
   Trithion
   Volck Oil

246. **TEXAS SAGE**
   Talstar

247. **TOLMIEA**
   Talstar

248. **TREES**
   (All kinds)
   Azodrin
   Bacillus
   Thuringiensis
   Baytex
   Creosote
   Diazinon
   Dibrom
   Dimilin
   Di-Syston
   Dursban
   Dylox
   Ethion
   Guthion
   Imidan
   Kelthane
   Kryocide
   Lindane
   Malathion
   Matacil
   Mavrik
   Mesurol
   Meta-Systox-R
   Methomyl
   Methoxychlor
   Methyl Parathion
   Morestan
   Nicotine
   Orthene
   Proxol
   Pydrin
   Sevin
   Sumithion
   Systox
   Talstar
   Tedion
   Thiodan
   Trithion
   Turcam

Volck Oil
Zectran

249. **TRUMPET-**
   FLOWER
   Trithion

250. **TULIPS**
   Chlordane
   Diazinon
   Dylox
   Guthion
   Malathion
   Meta-Systox-R
   Nicotine
   Thiodane

251. **TULIP TREE**
   Baytex (Entex)
   Diazinon
   Dylox
   Ethion
   Guthion
   Malathion
   Meta-Systox-R
   Thiodan

252. **TURF**
   (General)
   Affirm
   Amdro
   Aspon
   Baygon
   Cythion
   Dasanit
   Delnav (Deltic)
   Diazinon (Sarolex)
   Dibrom
   Doom
   Dursban
   Dyfonate
   Dylox
   Ethion
   Ficam-W
   Kelthane
   Lindane

Logic
Malathion
Mavrik
Mesurol
Methomyl
Mocap
Nemacur
Nexion
Oftanol
Orthene
Proxol
Pyrethrin
Sevin
Talstar
Telone
Thiodan
Trithion
Turcam
Vapam
Vorlex
Zectran

253. VELVET PLANT
Knox-Out

254. VERBENA
Baytex (Entex)
Diazinon
Dylox
Ficam-W
Guthion
Knox-Out
Malathion
Meta-Systox-R
Pounce
Talstar
Thiodan
Turcam

255. VIBURNUM
Azodrin
Baytex (Entex)
Dasanit
Diazinon
Dylox

Ethion
Guthion
Malathion
Meta-Systox-R
Methomyl
Nemacur
Nicotine
Oil
Orthene
Pentac
Talstar
Thiodan

256. VIOLETS
Delnav
Diazinon
Dylox
Endrin
Enstar
Guthion
Knox-Out
Kelthane
Malathion
Meta-Systox-R
Nicotine
Pyrethrins
Talstar
Thiodan
Trithion

257. VIRVIA
Knox-Out

258. WALLFLOWER
Rotenone

259. WANDERING JEW
Knox-Out
Pyrenone
Talstar

260. WILLOW
Diazinon
Dibrom
Dylox

Ethion
Guthion
Imidan
Malathion
Meta-Systox-R
Methomyl
Orthene
Thiodan

261. YARROW
Orthene

262. YEW (Taxus)
Baytex (Entex)
Cygon (Dimethoate)
Dasanit
Diazinon
Dibrom
Di-Syston
Dylox
Ethion
Ficam-W
Guthion
Imidan
Malathion
Meta-Systox-R
Methomyl
Mocap
Morestan
Nemacur
Orthene
Pentac
Talstar
Thiodan
Turcam

263. YUCCA
Knox-Out

264. ZINNIA
Baytex (Entex)
Diazinon
Dibrom
Di-Syston
Dylox
Ficam-W

Guthion
Knox-Out
Malathion
Meta-Systox-R
Methomyl
Nicotine
Orthene
Pentac
Pounce
Rotenone
Sevin
Talstar
Thiodan
Turcam

INSECTICIDES

ACEPHATE, ORTHENE
Mfg: Chevron

Ants, aphids, armyworms, bagworm, bean leaf beetle, birch leaf miners, blackgrass bugs, bollworms, boxelder bugs, budworms, brown cranberry spanworm, cockroaches, European corn borer, blackfly, cabbage looper, California oak moth, cankerworms, corn earworm, Cuban laural thrips, cutworms, Douglas fir tussock moth, earwigs, Eastern tent caterpillar, European corn borer, Elmleaf beetle, fall webworm, flea beetle, flea hoppers, forest tent caterpillar, grasshoppers, green bug, green clover worm, gypsy moth, hornworms, Japanese beetle, lace bugs, leafhoppers, leaf miners, leaf rollers, loopers, lygus, mealybugs, mirids, Mormon crickets, Nantucket pine tip moth, omnivorous leaf rollers, pillbug, pink bollworm, plant bugs, Ponderosa pineneedle miner, red-humped caterpillar, root weevils, sawflies, scales, silver spotted skiper, stinkbugs, sod webworms, three-cornered alfalfa hopper, thrips, tobacco bugworm, velvet bean caterpillar, wasps, webworms and whiteflies.

ALDICARB, TEMIK
Mfg: Union Carbide

Aphids, boll weevile, citrus mites, Colorado potato beetle, cotton leaf perforator, fleahoppers, leafhoppers, leaf miners, lygus, mealybugs, Mexican bean beetle, mites, nematodes, plant bugs, root maggots, three cornered alfalfa hopper, thrips, whiteflies.

ALUMINUM PHOSPHIDE, PHOSTOXIN, FUMITOXIN
Mfg: Degesch and Pestcon Systems

Stored grain insects.

AMINOCARB, MATACIL
Mfg: Mobay

Spruce budworm, jackpine budworm

AMITRAZ, BAAM, MITAC
Mfg: Nor-Am

Mites, pear psylla.

ASPON
Mfg: Stauffer

Cinch bugs, sod webworms.

AVERMECTRIN, AFFIRM, AVID
Mfg: MSD Ag Vet

Fire ants, 2-spotted mites, leafminers

25

AZINOPHOS-METHYL, GUTHION   *Mfg: Mobay*

Alfalfa plant bug, alfalfa weevil, American plum borer, aphids, apple maggot, blueberry maggot, boll weevil, bollworm, brown cotton leaf worms, cabbage maggot, carrot weevil, cherry fruit fly, cherry leafminer, citrus thrips, clover leaf weevil, codling moth, Colorado potato beetle, cone midge, cone moth, cone worm, cotton leafhopper, cotton leafworm, current borer, diamondback moth, European apple sawfly, European brown snail, European fruit lecanium, European pine shoot moth, European red mite, eye-spotted bud moth, false webworm, flea beetle, fleahoppers, filbert worm, Forbes scale, fruit flies, fruit tree leaf roller, fruitworms, grape berry moth, grape bud beetle, grape cane girdlers, hickory shuckworm, imported cabbageworm, hornworm, lacebugs, leafhoppers, leaf miners, leaf rollers, lesser clover leaf weevil, lesser peach twig borer, lygus, May beetles, mealy bug, Mexican bean beetle, Mineola moth, mites, Nantucket pine tip moth, navel orangeworm, oblique-banded leaf roller, orange tortrix, Oriental fruit moth, peachtree borer, peach twig borer, pecan case bearer, pink bollworm, plum curculio, putnam scale, rapid plant bug, raspberry root borer, red-banded leaf roller, scales, seedworms, Southern green stinkbug, spittlebug, stinkbug, strawberry leaf roller, tarnished plant bug, thrips, tip worm, tobacco budworm, tobacco hornworm, tobacco flea beetle, tuber moth, Tussock moth, twig girdlers, webworms.

BACILLUS-THURINGIENSIS,   *Mfg: Abbott, Nor-Am,*
BACTISPENE, BACTUR, BIOTROL, BTB,   *and Sandoz*
DIPEL, JAVELIN, SOK-BT, THURICIDE

Alfalfa caterpillar, almond moth, Amorbia moth, artichoke plume moth, bagworm, banana skipper, bollworm, bud worms, Cabbage looper, California oak moth, cankermoths, cankerworm, celery leaf tier, celery looper, cutworms, diamond-back moth, difoliatry caterpillar, Douglas fir tussock moth, elm spanworm, European corn borer, fall webworm, filbert leaf roller, forest tent caterpillar, fruit tree leaf roller, grape leaf folder, Great Basin tent caterpillar, green clover worm, gypsy moth, hornworms, imported cabbageworm, Indian meal moth, loopers, oakmoth, olecinder moth, omnivorous leaf roller, omnivorous looper, orange dog, podworms, range caterpillar, red-banded leaf roller, red-humped caterpillar, rind worm, tent caterpillar, salt marsh caterpillar, sorghum, headworm, soybean looper, spruce budworm, sunflower moth, tobacco budworm, tobacco hornworm, tobacco moth, tomato fruitworm, tufted apple bud moth, varigated leaf roller, velvet bean caterpillar, wax moth larvae.

BENDIOCARB, FICAM,
TURCAM   *Mfg: Nor-Am Chemicals*

Ants, aphids, Azalea caterpillar, Azalea leaf miner, bagworms, bees, billbugs, Black-turf grass, black vine weevil, bronze birch borer, carpet beetles, centipedes, chinch bugs, citrus blackfly, clothes moth, cockroaches, crickets, drugstore beetle, earwigs, elm leaf beetle, European chafer, fireants, firebrats, fleas, flies, flour beetles, grain beetles, green June bugs, ground beetles, gypsy moth, Japanese beetle, lacebugs, leafhoppers, lesser grain borer, mealybugs, millipedes, mole-crickets, mosquitoes, Nantucket pinetip moth, Northern mashed chafer, Nothern pine weevil, obscure root weevil, oleander caterpillar, Pales pine weevil, pantry pests, pillbugs, pine spittlebug, pine tip moth, poplar tentmakers, rice weevils, scales, scorpions, silverfish, sowbugs, spider beetles, spiders, spring cankerworm, tent caterpillars, termites, thrips, ticks, tobacco moth, wasps, webworms, white grub, willow leafbeetle, whiteflies, Yellow-necked caterpillar.

BIFENTHRIN, BRIGADE,
CAPTURE, TALSTAR   *Mfg: FMC*

Aphids, armyworms, mealy bugs, omnivorous leafrollers, scales, two spotted spider mite, white flies.

BORIC ACID   *Mfg: Numerous*

Cockroaches.

BROMOPHOS, NEXION   *Mfg: Celamerck and EM*
                                                   Labs

Chinch bugs.

BTI, BACTIMOS, TEKNAR,
VECTOBAC   *Mfg: Abbott, Sandoz,*

Blackflies, mosquitoes.

CALCIUM CYANIDE   *Mfg: Degesch American*

Aphids, mites, mushroom flies, rodents, spring tails, stored grain insects, thrips, white flies, yellow jackets.

CARBARYL, SEVIN   *Mfg: Union Carbide*

Alfalfa caterpillar, alfalfa looper, alfalfa weevil, ants, Apache cicada, aphids, apple maggot, apple mealybug, apple rust mite, apple sucker, armyworms, asparagus beetles, aster leafhopper, bagworm, bean leaf beetle, bean leaf roller, bedbugs, bees, birch leaf miner, black cherry aphid, black scale, blister beetles, blueberry maggot, bluegrass billbug, body lice, boll weevil, bollworm, boxelder bug, boxwood leaf miner, brown dog tick, brown soft scale, cabbage looper, calico scale, California red scale, cankerworms, catfacing insect, cherry fruit fly, cherry fruitworm, chicken mite, chiggers, chinch bugs, citricola scale, citrus cutworm, climbing cutworm, clover head weevil, cockroaches, codling moth, Colorado potato beetle, corn earworm, corn rootworm, cotton fleahopper, cotton leaf perforator, cowpea curculio, crab louse, cranberry fruitworm, crickets, cucumber beetle, cutworms, darkling ground beetles, diamondback moth, earwigs, Eastern apple sawfly, Eastern spruce gall aphid, Eastern tent caterpillar, elm leaf beetle, eriophyid mites, European apple sawfly, European chafer, European corn borer, European earwig, European fruit lecanium, eye-spotted bud moth, fall armyworm, fall cankerworm, filbert leaf roller, filbert moth, filbert worm, fireworms, flea beetle, fleas, Forbes scale, forest tent caterpillar, fowl tick, frosted scale, fruit tree leaf roller, fruitworms, grape berry moth, grape leaf folder, grape leaf hopper, grape leaf skeletonizer, grasshoppers, great basin tent caterpillar, green clover worm, green fruitworms, green June beetle, green June bug, gypsy moth, Harlequin bug, head lice, hornets, hornworm, housefly, imported cabbageworm, imported fire ant, Japanese beetle, June beetles, lace bugs, leafhoppers, leaf rollers, lecanium scales, Lepidopterous larvae, lesser peach tree borer, lice, lima bean pot borers, lygus bugs, maple leaf cutter, meadow spittlebug, mealybugs, mealy plum aphid, melonworm, millipedes, mimosa webworm, Mexican bean beetle, mole crickets, mosquitoes, moth flies, navel orangeworm, Northern fowl mite, oak leaf miners, olive scale, omnivorous leaf tier, omnivorous leaf roller, orange tortrix, Oriental fruit moth, oyster shell scale, pandemis moth, peach twig borer, pear leaf blister moth, pear psylla, pear rust mite, pickleworm, pink bollworm, pinworms, plant bugs, plum curculio, prune leafhopper, psyllids, raspberry aphids, rebanded leaf roller, rosebug, salt marsh caterpillar, San Jose scale, sand flies, sap beetles, saw flies, scales, six-spotted leafhopper, snowy tree crickets, sod webworm, sorghum midge, sowbugs, spittlebugs, spruce budworm, squash bug, stem beetles, stinkbugs, strawberry leaf roller, strawberry weevil, striped blister beetle, sunflower beetle, tadpole shrimp, tarnished plant bug, tentiform leaf miner, thornbug, three-cornered alfalfa hopper, thrips, ticks, tobacco budworm, tobacco flea beetle, tomato fruitworm, tomato hornworm, Tussock moth, vegetable weevil, velvet bean caterpillar, wasps, webworms, Western bean cutworm, Western Tussock moth, white apple leafhopper, willow leaf beetle, yellow scale.

CARBOFURAN, FURADAN *Mfg: FMC and Mobay*

Alfalfa blotch leaf miner, alfalfa snout beetle, alfalfa weevil, aphids, armyworms, banana root borer, banks grass mite, cereal leaf beetle, chinch bugs, clearwinged bug, clearwinged borer, Colorado potato beetle, cone beetles, cone worms, cone borers, cottonwood leaf beetle, cottonwood twig borer, corn rootworms, cutworms, elm leaf beetle, European corn borer, flea beetles, grape phylloxera, grasshoppers, greenbugs, hornworms, lygus bugs, Mexican bean beetle, mosquitoes, nematodes, Pales Weevil, pitch eating weevil, potato leaf-hopper, potato tuberworm, rice water weevil, root weevil, seed bugs, seed corn maggot, Southern corn rootworm, Southwestern corn borer, stalk rot, sugar beet maggots, sugarcane borer, sunflower beetle, sunflower weevil, thrips, tobacco budworm, tuberworm, tung borer, white grubs, wireworms.

CARBOPHENOTHION, TRITHION *Mfg: Stauffer*

Aphids, bagworms, blueberry maggot, chinch bug, codling moth, corn root-worm, cotton leafworm, flea hoppers, greenbug, green clover worm, green stinkbug, leaf miners, leaf perforator, leafhoppers, mealybugs, Mexican bean beetle, mites, onion maggot, pear psylla, potato psylla, scales, sorghum midge, spittlebug, thirps, twig borer.

CHLORDANE *Mfg: Velsicol*

Termites.

CHLORDINEFORM, FUNDAL, GALECRON *Mfg: CIBA-Geigy and Nor-Am*

Bollworm and tobacco budworm.

CHLOROBENZILATE, ACARABEN *Mfg: CIBA-Geigy*

Mites.

CHLORPYRIFOS-METHYL, RELDAN *Mfg: Dow, Gustafson*

Angoumoirs grain moth, confused flour beetle, flour beetles, granary weevils, Indian meal moths, lesser grain beetle, lesser grain borers, rice weevils, saw toothed beetles.

CHLORPYRIFOS, DURSBAN, LORSBAN *Mfg: Dow, Gustafson*

Adelgibs, ants, aphids, apple maggot, armyworms, ash borer, bagworms, banded sunflower moth, beetles, billbugs, bollworms, boll weevil, borers, boxelder bugs, brown dog ticks, carpet beetles, catalpa sphinx, centipedes, chiggers, chinch bugs, cockroaches, codling moth, conkerworms, corn borers, corn earworm, corn rootworm, cotton wood beetles, cotton leaf perforator, cranberry span worm, cranberry weevil, crickets, cutworms, dogwood borer, earwigs, Eastern tent caterpillars, Egyption alfalfa weevil, elm bark beetle, elm span worm, European apply sawfly, European chafer, European corn borer, European crane fly, eyespotted budmoth, fall webworm, fiery skyger, firebrats, fireworms, fleabeetle, fleahopper, fleas, fruit leafroller, gnats, grasshoppers, green bugs, green fruitworm, hickory shuckworm, hornets,hyperodes weevil, Indian meal moth, Japanese beetle, jackpine budworm, June beetle, Juniper webworm, katydids, keds, lacebugs, leaf cutters, leafhoppers, leaf miners, leafrollers, lesser apple worm, lesser cornstalk borer, lesser peach tree borer, lilac moth, lygus, Mahogany webworm, maple leaf cutter, mealybugs, Mediterranean flour moth, millipedes, mimosa webworm, mint root borer, mirids, mites, mosquitoes, moths, naval orange worms, needle miners, oakworms oak skeletonizers, oblique banded leafroller, oleander caterpillar, onion maggots, orange stripped oakworm, orange tortrix, oriental fruit moth, pales beetles, pandomis leafroller, peach tree borer, peach twig borer, pecan nut casebearer, pecan weevil, periodical cicada, phylloiera, pink bollworm, pitch eating weevil, plant bugs, plum curculio, poplar tentmaker, puss caterpillar, red banded leafroller, red flour weevil, red humped caterpillar, rhodonendron borer, rice weevils, root maggots, rose chafer, salt marsh caterpillar, San Jose scale, saw flies, sawtoothed grain beetle, scales, scorpions, silverfish, sod webworms, Southern mashed chafer, Southern pine beetles, Southwestern corn borer, sawbugs, spargnothis fruit worms, spiders, spittlebugs, spring elm caterpillars, springtails, spruce budworm, sugar beet root maggots, symphylons, tarnished plant bug, tent caterpillars, termites, thornbug, thrips, ticks, tobacco budworms, tufted apple budmoth, varigated leafroller, vegetable weevils, Walnut caterpillar, Walnut scale, wasps, weevils, western tussock moth, whiteflies, white grubs, wireworms wolly bears, yellow jackets, yellow necked caterpillar.

COUMAPHOS, CO-RAL　　　　　　　*Mfg: Mobay*

Fleeceworms, grubs, horn flies, keds, lice, Northern fowl mite, poultry red mite, screwworms, ticks, face flies, scab mite, wool maggots.

CROTOXYPHOS, CIODRIN　　　　　　*Mfg: Numerous*

Face flies, horn flies, houseflies, lice, lone star ticks, stable flies, winter ticks.

CYHEXATIN, PLICTRAN　　　　　　　*Mfg: Dow*

Mites.

CYPERMETHRIN, AMMO, DEMON, CYMBUSH

Mfg: ICI Americas and FMC

Beet armyworm, boll weevil, cabbage looper, cotton bollworm, cotton leafhopper, cotton leaf perforator, cutworms, fall armyworms, lygus bugs, pink bollworm, plant bugs, soybean thrips, tarnished plant bugs, tobacco budworm, tobacco thrips, whiteflies, yellow striped armyworm.

CYROMAZINE, LARVADEX, TRIGARD

Mfg: CIBA-Geigy

Flies, leafminers.

DEMETON, SYSTOX

Mfg: Mobay

Aphids, foliar nematodes, lace bugs, leafhoppers, leaf miners, mealybugs, mites, scales, thrips, whiteflies.

DIAZINON, KNOX-OUT

Mfg: CIBA-Geigy, Pennwalt

Alfalfa weevil, ants, apple aphid, apple maggot, bagworms, banded cucumber beetles, bean aphid, Bermuda mite, billbugs, black cherry aphid, blackheaded fireworms, black scale, brown soft scale, cabbage maggot, carnation bud mite, carpet beetles, carrot rust flies, cherry fruit flies, cherry rust mites, chiggers, chinch bug, citrus aphid, clover mite, codling moth, Colorado potato beetle, corn earworm, corn rootworm, corn sap beetle, cottony cushion scale, cotton leaf perforator, cotton leafworm, cranberry fruitworm, cranberry girdler, crickets, cutworms, cyclamen mite, diamondback moth, dipterous leaf miner, earwigs, European fruit lecanium, European red mite, eye-spotted bud moth, fall armyworm, filbert leaf roller, flea beetles, fleas, flies, Forbes scale, fruit tree leaf roller, grape berry moth, grape leaf folder, grasshoppers, green apple aphid, Harlequin bug. Holly bud moth, houseflies, imported cabbageworm, keds, lawn billbug, leaf curl plum aphid, Mexican bean beetle, Mimosa webworm, mites, olivescale, onion thrips, onion maggot, Oriental fruit moth, Pacific spider mite, Parlotoria scale, pea aphid, peach twig borer, pear leaf blister mite, pear psylla, pecan nut casebearers, plant bug, privet mite, raspberry root borer, rosy apple aphid, San Jose scale, sciarids, scorpions, seed corn maggot, serpentine leaf miner, silverfish, sod webworm, sorghum midge, Southern armyworm, Southern corn rootworm, Southwestern corn borer, sowbugs, spiders, spittlebug, spotted alfalfa aphid, springtails, strawberry leaf roller, tentiform leaf miner, thrips, ticks, two-spotted mite, vinegar flies, walnut caterpillars, walnut scale, whiteflies, white grubs, wireworms, hyperodes weevil, woolly apple aphid, yellow clover aphid, yellow jackets.

DICHLOROVOS, DDVP, VAPONA   *Mfg: DuPont*

Ants, aphids, bedbugs, boulder bugs, brown dog ticks, carpet beetle, centipedes, cigarette beetles, clothes moths, clover mites, cockroaches, crickets, dried fruit beetle, fleas, flies, flying moths, fruit flies, gnats, hornets, lice, mealybugs, mites, millipedes, mosquitoes, phorid flies, roaches, sciarid flies, scorpions, silverfish, sowbugs, spiders, ticks, tobacco moths, wasps, whiteflies.

DICOFOL, KELTHANE   *Mfg: Rohm & Haas*

Almond mite, apple rust mite, brown mite, carmine mite, citrus flat mite, citrus red mite, citrus rust mite, clover mite, desert mite, European red mite, grass mite, McDaniel mite, Pacific mite, peach silver mite, plum nursery mite, schoene mite, six-spotted mite, strawberry mite, tomato russet mite, tropical russet mite, two-spotted mite, Williamette mite, yellow mite, Yuma mite.

DICROTOPHOS, BIDRIN   *Mfg: DuPont*

Aphids, bagworms, black flea hopper, boll weevil, camellia scale, Colorado potato beetle, cotton flea hopper, elm leaf beetle, European elm bark beetle, European elm scale, European pine sawfly, grasshoppers, leafhoppers, leaf perforators, lygus bugs, Mexican bean beetle, Mimosa webworm, pine sawfly, pine spittlebug, pine tip moths, potato flea beetle, salt marsh caterpillar, scales, southern pine beetle, spider mites, stinkbugs, taxus mealybugs, tea scale, thrips, treehoppers, whiteflies.

DIENOCHLOR, PENTAC   *Mfg: Sandoz*

Mites.

DIFLUBENZURON, DIMILIN   *Mfg: Uniroyal*

Armyworms, boll weevil, Douglas fir tussock moth, forest tent caterpillars, green cloverworm, Gypsy moth, Mexican bean beetle, mosquitoes, Nantucket pine tipmoth, sciarid fly, spruce budworm, velvet bean caterpillar.

DIMETHOATE, CYGON, DE-FEND,   *Mfg: Numerous*
REBELATE

Alfalfa hopper, alfalfa weevils, aphids, bagworms, banks grass mite, bean beetle, bean leaf beetle, corn rootworms, Douglas fir cone midge, European pine shoot moth, flea hoppers, grape leafhoppers, grasshoppers, houseflies, iris borer, lace bugs, leafhoppers, leaf miners, Loblolly pine sawfly, lygus bugs, Mexican bean beetle, midges, mites, Nantucket pine tip moth, pear psylla, pepper maggot, plant bugs, scales, thrips, whiteflies, Zimmerman pine moth.

DIMETHYNONANE, **PRO-DRONE**   *Mfg: Stauffer*

Imported red fire ants.

DINOCAP, **KARATHANE**   *Mfg: Rohm & Haas*

Citrus red mite, clover mite, European red mite, peach silver mite, plum nursery mite, powdery mildew, red spider mite, two-spotted mite.

DIOXATHION, **DELTIC**   *Mfg: Nor-Am Ag Chemicals*

Ants, chiggers, clover mite, crickets, European red mite, fleas, harvestmen, leafhoppers, spruce spider mite, ticks, 2-spotted mite.

DISULFOTON, **DI-SYSTON**   *Mfg: Mobay*

Alfalfa, aphids, banks grass mite, birch tree miner, brown soft scale, Colorado potato beetle, elm leaf beetle, flea beetle, grasshoppers, Hessian fly, Holly leaf miner, lace bug, leafhoppers, leaf miners, mealybug, Mexican bean beetle, Mimosa webworm, mites, pine tip moth, potato psylid, root maggot, rootworms, scales, sorghum midge, Southern potato wireworm, thrips, whiteflies.

DNBP, **ELGETOL**   *Mfg: FMC Corp.*

Mites, peach twig borer, scales, aphids, leaf crumpler, California casebearer, eye-spotted bud moth, pecan phylloxera, European fruit lecanium.

ELCAR   *Mfg: Sandoz*

Corn earworm, cotton bollworm, sorghum head worm, soybean pod worm, tobacco budworm, tomato fruitworm, Heliothis Spp.

ENDOSULFAN, **THIODAN**   *Mfg: FMC*

Aphids, army cutworm, armyworms, artichoke plume moth, banded cucumber beetle, bark beetles (scolytid sp), bean leaf beetle, bean leaf skeletonizer, black vine weevil, blister beetle, boll weevil, bollworm, boxelder bug, cabbage looper, catfacing insects, cereal leaf beetle, Colorado potato beetle, consperse stinkbug, corn earworm, cotton leaf perforator, cowpea curculio, cross-striped cabbage-worm, cucumber beetles, cutworms, diamondback moth larvae, dogwood borer, European corn borer, eye-spotted bud moth, false chinch bug, filbert leaf roller, flea beetles, fruit tree leaf roller, garden symphylan, grape leafhopper, green June bug larvae, green stinkbug, Harlequin bug, hornworms, imported cabbage-worm, iris borer, leaf-footed bug, leaf miner, leatherleaf fern borer, lesser peach tree borer, lilac borer, lygus bugs, meadow spittlebug, melon leafhopper, melon leaf miner, melonworm, Mexican bean beetle, mites, omnivorous leaf roller, other mirids, pea weevil, peach tree borer, peach twig borer, pear psylla, pecan nut case bearer, pillbugs, plant bugs, potato flea beetles, potato leafhopper, potato psyllid, potato tuberworm, rindworm, rose chafer, serpentine leafminer, Southern green stinkbug, squash beetle, squash bug, squash vine borer, stinkbugs, striped flea beetle, sugar beet webworm, sugarcane borer, sunflower moth, sweet potato flea beetle, tarnished plant bug, three-lined potato beetle, thrips, tobacco budworm, tobacco flea beetle, tobacco hornworm, tomato fruitworm, tomato hornworm, webworm, Western bean cutworm, whiteflies, wood borers, yellow-striped armyworm, Zimmerman pine moth.

EPN   *Mfg: Sandoz*

Aphids, beet armyworm, boll weevil, bollworm, bud moth, citrus thrips, codling moth, corn earworm, cotton fleahopper, cotton leafworm, European red mite, eye-spotted bud moth, fall armyworm, fall webworm, flea beetles, fruit tree leaf miners, lecanium scale, May beetles, Mexican bean beetle, mosquitoes, olive scale, orange tortrix, Oriental fruit moth, peach tree borer, pear psylla, pecan nut casebearer, pecan twig girdler, pecan weevil, plum curculio, potato aphid, purple mite, red spider mite, russet mite, scales, Southern green stinkbug, stinkbug, sugar beet webworm, tobacco budworm, tomato fruitworm, two-spotted mite, walnut aphid, walnut caterpillar, Williamette mite, yellow striped armyworm.

ETHION   *Mfg: FMC*

Aphids, bean leaf beetle, chinch bug, citrus thrips, citrus whitefly, codling moth, cotton leafworm, cucumber beetles, European fruit lecanium, grape berry moth, leafhoppers, leaf miner, lygus bugs, melon leaf miner, Mexican bean beetle, millipeds, mites, onion maggot, other mirids, pear psylla, potato leafhopper, scales, seed corn maggot, serpentine leaf miner, sod webworm, sorghum midge, strawberry leaf roller, taxus mealybug, thrips, walnut husk fly.

ETHOPROPHOS, MOCAP   *Mfg: Rhone Poulenc*

Black turfgrass ateanius, bluegrass billbug, chinch bugs, corn rootworms, cutworms, European chafer, flea beetle, garden symphylans, Japanese beetle, nematodes, sod webworms, wireworms.

FAMPHUR, WARBEX   *Mfg: American Cyanamid*

Cattle grubs, lice.

FENITROTHION, SUMITHION   *Mfg: Stauffer, Sumitomo*

Cockroaches, mosquitoes, Southern pine beetles, spruce budworm.

FENOXYCARB, LOGIC   *Mfg: Maag Agrochemicals*

Fireants.

FENSULFOTHION, DASANIT   *Mfg: Mobay*

Aphids, corn rootworms, flea beetle, lesser cornstalk borer, nematodes, onion maggot, wireworms, seed corn beetle, seet corn maggot, thirps.

FENTHION, BAYTEX   *Mfg: Mobay*

Ants, aphids, armyworms, bagworms, cadelle, cattle grubs, cockroaches, confused flour beetle, crickets, fleas, flour moths, grain moth, heartworms, hookworms, horn fly, housefly, leafhoppers, lice, meal moth, midges, mites, mosquitoes, moths, saw-toothed grain beetle, scale, ticks, vinegar flies, wasps, water bugs.

FENVALERATE, PYDRIN, PYRID   *Mfg: DuPont*

Alfalfa looper, ants, aphids, apple maggot, armyworm, artichoke moth, banded sunflower moth, bean leafbeetle, black cutworm, boll weevil, cabbage looper, cadelles, carpet beetles, carrot weevil, centipedes, cherry fruit fly, chinch bugs, codling moth, cigarette beetles, cockroaches, Colorado potato beetle, conebugs, confused flour beetle, corn earworm, corn rootworm, cotton bollworm, cotton leaf perforator, crickets, cucumber beetle, cutworms, cowpea curculio, diamondback moth, drug store beetle, European corn borer, flea beetles, fleas, flies, form leafminers, fruitworm, granary weevil, grasshoppers, green clover worm, Heliothis Spp., hickory shuckworm, hornworm, imported cabbage worm, leafhoppers, leaf miners, leaf rollers, lesser appleworm, lesser peach tree borer, lice, lygus, meal moth, Mexican bean beetle, Nantucket pine tip moth, Naval orange worm, Oriental fruit moth, pea leaf weevil, peach twig borer, pear psylla, pear rust mite, pear slug, pecan casebearer, pecan weevil, periodical cicados, pickleworm, pinworm, pink bollworm, plum borer, plum curculio, potato psyllid, potato tuberworm, red blanket leaf roller, red-necked peanut worm, rindworm, rust red flour beetle, salt marsh caterpillar, San Jose scale, saw-toothed grain beetle, scales, seed bugs, seed weevils, sheep keds, silverfish, Southern green stinkbug, Southwestern cornborer, sow bugs, spiders, squash bug, surgarcane borer, sunflower moth, sunflower stem weevil, tarnished plant bug, tent, ticks, tobacco budworm, tufted apple budworm, vegetable leaf miner, Velvet bean caterpillar, weevils, Western bean cutworm, whiteflies.

FLUCYTHRINATE, PAY-OFF
Mfg: DuPont

Aphids, cabbage looper, cotton bollworm, cotton leaf perforator, boll weevil, flies, leaf miners, leaf rollers, lygus, oblique-banded leafroller, pink bollworm, plum curculio, salt marsh caterpillar, spotted tentform leafminer, tobacco budworm.

FLUVALINATE, MAVRIK, SPUR
Mfg: Sandoz

Aphids, armyworms, boll weevils, bollworm, corn earworm, cotton leaf perforator, cucumber beetle, cutworm, diamond back moth, flea beetles, flea hoppers, fleas, grasshoppers, imported cabbage worm, Japanese beetle, leaf feeding caterpillars, leafhoppers, leaf miners, leaf rollers, loopers, lygus, mites, oriental fruit moth, peach twig borer, pear psylla, root weevils, salt marsh caterpillar, scales, thrips, ticks, tobacco budworm, webworms, whiteflies.

FONOFOS, DYFONATE
Mfg: Stauffer

Aphids, Cabbage maggot, chinch bugs, corn rootworms, cutworms, European corn borer, garden symphylans, Gelechiid moth, greenbugs, lesser cornstalk borer, onion maggot, peanut rootworm, seed corn beetle, seed corn maggot, sod webworm, sugar beet root maggot, symphylans, white grub, wireworms.

FORMETANATE, CARZOL
Mfg: Nor-Am Chemicals

Apopka weevil, consperse stinkbug, lygus bug, mites, tentiform leafminers, thrips.

HEPTACHLOR
Mfg: Velsicol

Ants.

HEXAKIS, VENDEX
Mfg: DuPont

Mites.

ISOFENPHOS, AMAZE, OFTANOL
Mfg: Mobay

Asiatic garden beetle, ataenius, billbugs, chafers, chinch bugs, corn rootworm, craneflies, green June Beetle, hyperodes weevil, Japanese beetle, mole crickets, Oriental beetle, sod webworm, termites, white grubs.

KINOPRENE, ENSTAR
Mfg: Sandoz

Aphids, mealybugs, fungus gnats, scales, whiteflies.

LINDANE

Mfg: Numerous

Ants, aphids, apple grain aphid, armyworms, bean weevil, bedbugs, cabbage maggots, carpet bugs, chiggers, climbing cutworms, clothes moth, confused flour beetle, crickets, cucumber beetle, curculio, Diabrotica, European sawfly, false wireworms, flea beetles, fleas, fleeceworms, fowl tick, Fuller rose weevil, garden symphyhids, gladiolus thrips, gnats, granary weevil, grasshopper, houseflies, Indian meal moth, keds, lace bugs, leaf miners, lice, locust borer, loopers, lygus bugs, mange, melonworms, mosquitoes, mushroom flies, onion maggot, pecan curculio, pecan phylloxera, phylloxera, pickleworms, pine bark aphids, pine root collar beetle, powder post beetle, psyllid, rice weevil, roaches, rose midge, rosy apple aphid, saw-toothed grain beetle, screwworm, seed corn maggot, silverfish, spiders, spittlebugs, spruce gall aphid, squash vine borer, stable flies, strawberry weevil, tarnished plant bug, termites, thrips, ticks, webworms, whitefly, white grubs, white pine weevil, wireworms, woolly apple aphids.

MAGNESIUM PHOSPHIDE, FUNI-CEL

Mfg: Degesch and Pestcon Systems

Almond moth, Angoumois grain moth, bean weevil, cadelle, cereal leaf beetle, cigarette beetle, confused flour beetle, dernested beetle, dried fruit beetle, dried fruit moth, European grain moth, flat grain beetle, fruit flies, granary weevil, Indian meal moth, lesser grain borer, Mediterranean flour moth, pink bollworm, raisin moth, red flour beetle, rice weevil, rusty grain beetle, saw-toothed grain beetle, tobacco moth.

MALATHION, CYTHION

Mfg: American Cyanamid

Alfalfa weevil, aphids, armyworms, bagworms, beet leafhopper, black cherry aphid, blackheaded fireworm, black scale, blueberry maggot, boll weevil, bud moth, cabbage looper, cereal leaf beetle, cherry fruit fly, chicken red mite, clover mite, codling moth, confused flour beetle, corn rootworm, cranberry fruitworm, cucumber beetle, curculio, drosophila, European red mite, false chinch bug, flat grain beetle, field chickets, flies, Florida red scale, Forbes scale, fruit tree leaf roller, granary weevil, grasshoppers, green apple aphid, horn flies, imported cabbage worm, Indian meal moth, Japanese beetle, juniper scale, keds, lace bugs, leafhoppers, lesser grain borer, lice, lygus, mealybugs, mites, Monterey pine scale, mosquitoes, Northern fowl mite, onion maggot, Oriental fruit moth, peach tree borer, pear psylla, pickleworm, pine needle scale, plum curculio, potato aphid, potato leafhopper, poultry lice, poultry ticks, purple scale, red-banded leaf roller, red flour beetle, rice stinkbug, rice weevil, rose scale, rosy apple aphid, rusty grain beetle, saw-toothed grain beetle, soft brown scale, soft scale, spittlebug, squash vine borer, strawberry leaf roller, strawberry root weevil, tarnished plant bugs, tent caterpillars, thirps, ticks, two-spotted mites, unspotted tentiform leaf miner, vetch bruchids, whiteflies, Williamette mite, woolly apple aphid, yellow-backed caterpillars, yellow scale.

METALDEHYDE   Mfg: Numerous

Slugs and snails.

METHAMIDOPHOS, MONITOR   Mfg: Mobay and Chevron

Aphids, beet armyworm, cabbage looper, Colorado potato beetle, cutworms, diamondback moth, flea beetles, Imported cabbageworm, Leafhoppers, mites, potato tuberworm, stinkbugs, thirps.

METHIDATHION, SUPRACIDE   Mfg: CIBA-Geigy

Alfalfa Weevil, aphids, boll weevil, banksgrass mite, bollworms, budworms, citrus black fly, codling moth, Egyptian alfalfa weevil, flea beetles, greenbug, hickory shuckworm, hornworms, leafhoppers, leafworms, lygus, peach twig borer, scales, spider mites, spittlebugs, whiteflies.

METHIOCARB, MESUROL   Mfg: Mobay

Aphids, blueberry maggot, cherry fruit fly, mites, plum curulio, slugs, snails. Also used as a bird repellent.

METHOMYL, LANNATE, NUDRIN   Mfg: DuPont

Alfalfa caterpillar, alfalfa looper, alfalfa weevil, aphids, apple worms, armyworm, beet webworm, bollworm, budworm, cabbage looper, cabbage-worm, codling moth, corn earworm, cotton leaf perforator, cutworms, diamond-back moth, European corn borer, fall armyworm, flea beetles, grape berry moth, grape leaf holder, grape leaf rollers, green clover worm, green fruitworm, imported cabbageworm, leafhoppers, leaf miners, lygus bugs, melon worms, Mexican bean beetle, omnivorous leaf roller, orange tortix, Oriental fruit moth, picnic beetle, pickleworm, pinworms, potato tuberworm, red-banded leaf roller, salt marsh caterpillar, sod webworm, soybean looper, thirps, tobacco budworm, tobacco hornworm, tomato fruitworm, twig borers, velvet bean caterpillar, tufted apple budmoth, fruit tree leaf roller, spruce budworm, sorghum midge, Western tussock moth, orange tortix, leaf perforator, asparagus beetle, tarnished plant bug.

METHOPRENE, ALTOSID, APEX, DIACON, DIANEX, KABAT, MINE

Mfg: Sandoz

Almond Moth, cigarette beetles, confused flour beetle, fleas, flies, Indian meal moths, leaf miners, lesser grain borer, merchant grain beetle, mosquitoes, Pharoah's ant, red flour beetle, saw-toothed grain borer, Sciarid beetles, tobacco moth.

METHOXYCHLOR

Mfg: Kincaid Mfg.

Alfalfa caterpillar, alfalfa weevil, apple maggot, armyworms, asparagus beetle, bean leaf beetle, blister beetle, cabbageworm, Cadella flat grain beetle, canker-worm, cherry fruit fly, cherry fruit worm, codling moth, clover leaf weevil, Colorado potato beetle, confused flour beetle, corn earworm, cowpea curculio, cranberry fruitworm, cucumber beetle, fall armyworm, flat grain beetle, flea beetle, fleas, foreign grain beetle, gnats, grape berry moth, grape leaf skeleton-izer, horn flies, houseflies, Japanese beetle, keds, leafhoppers, lesser grain borer, lice, long-headed beetle, melonworm, Mexican bean beetle, mosquitoes, pear slug, pea weevil, plum curculio, red flour beetle, rice weevil, roaches, rose chafer, rose slugs, San Jose scale, saw-toothed grain beetle, soybean caterpillar, spittlebug, squash vine borer, stable flies, strawberry weevil, tent caterpillars, ticks, velvet bean caterpillar, webworm, wasps.

METHYL BROMIDE

Mfg: Dow and others

Stored grain insects, nematodes, commodity storage insects, soil insects, fungi and weeds.

METHYL PARATHION

Mfg: Monsanto

Alfalfa caterpillar, alfalfa seed chalcid, alfalfa weevil, aphids, armyworms, artichoke plume moth, bean leaf beetle, black grass bug, blister beetle, boll weevil, bollworm, chinch bug, clover leaf weevil, clover seed chalcid, codling moth, corn earworm, corn rootworm, cotton leaf worm, cotton leaf perforator, cowpea curculio, cutworms (climbing), Egyptian alfalfa weevil, European pine shoot moth, false chinch bug, flea beetles, fleahoppers, grape leaf folders, grasshoppers, green clover worm, green June beetle, imported cabbageworm, leafhoppers, leaf miners, leaf rollers, loopers (cabbage), lygus bugs, Mexican bean beetle, mites, mosquitoes (larvae), Nantucket pine tip moth, Oriental fruit moth, peach tree borer, plant bugs, plum curculio, potato psyllid, rice leaf miners, rice stinkbug, salt marsh caterpillar, scales, seed corn maggot, sorghum midge, stinkbug, tadpole shrimp, three-cornered alfalfa hopper, thrips, velvet bean caterpillar, vetch bruchid, webworms.

MEVINPHOS, PHOSDRIN   Mfg: DuPont

Alfalfa caterpillar, alfalfa weevil, aphids, artichoke plume moth, armyworms, asparagus beetle, cabbage looper, citrus cutworm, corn earworm, Dipterous leaf miners, cutworms, false chinch bugs, filter flies, fruit tree leaf roller, grape leaf folder, grasshoppers, grass worms, greenhouse whitefly, green stinkbugs, imported cabbageworms, leafhopper, leaf folder, leaf roller, lygus bugs, Mexican bean beetle, mites, omnivorous looper, orange tortrix, pink scavenger, diamondback moth, pickleworms, plant bugs, caterpillar, plum moth, psycodid flies, red-banded leaf roller, salt marsh caterpillar, strawberry leaf roller, thrips, tobacco budworm, Tussock moth, varigated cutworm, velvet bean caterpillar, webworms, western moth, tomato fruitworm, whiteflies.

MEXACARBATE, ZECTRAN   Mfg: Union Carbide

Aphids, armyworms, bagworms, birch leaf miner, cutworms, European pine shoot moth, flea beetle, Genista caterpillar, hackberry nipple gallmaker, holly budmoth, honey locust pod gall, Japanese beetle, leafhoppers, leaf miners, leaf rollers, loopers, lygus, mealybugs, millipedes, mites, needle miner, omnivorous leaftier, orange tortrix, plant bugs, rose chafer, scale crawlers, scales, slugs, snails, sod webworms, sowbugs, spiney elm caterpillar, spruce budworm, tent caterpillars, thrips, ugly nest caterpillar, whiteflies, yellow-necked caterpillars.

MONOCROTOPHS, AZODRIN   Mfg: DuPont

Aphids, beet armyworm, loopers, boll weevil, bollworms, Colorado potato beetle, cutworms, flea beetles, flea hoppers, leafhoppers, leaf miners, lesser cornstalk borer, loopers, lygus bugs, mites, pink bollworms, plant bugs, potato tuber worm, sawflies, sugarcane borer, thrips, tobacco hornworms, tobacco budworms, tomato fruitworm, whiteflies.

MYCAR   Mfg: Abbott Labs

Citrus rust mite

NALED, DIBROM   Mfg: Chevron

Alfalfa webworms, aphids, armyworms, blister beetles, bollworms, cabbage looper, California oakworm, cankerworms, citrus cutworm, clover mites, cockroaches, cotton leaf perforator, crane flies, cucumber beetle, diamondback moth, drosophila, earwigs, elm leaf beetle, fleas, flea beetles, flying moths, fruit flies, gnats, grasshoppers, hemlock borer, hornworms, houseflies, imported cabbageworms, leafhoppers, leaf miners, leaf perforator, leaf roller, loopers, lygus bugs, mealybugs, midges, mosquitoes, moths, mites, Oak webworm, omnivorous leaf

40

tier, orange tortrix, Oriental fruit moth, peach twig borer, pickleworms, range caterpillar, red spider mite, salt marsh caterpillar, sap beetle, strawberry leaf roller, Sierra fir borer, soft brown scale crawlers, spider mites, spittlebugs, stable flies, stinkbugs, tent caterpillars, thrips, ticks, tomato fruit flies, Tussock moth, webworms, whiteflies, Zimmerman pine moth.

NICOTINE SULFATE   Mfg: Black Leaf Products

Ants, aphids, apple red bugs, asparagus beetle, blackheaded fireworm, bud moth, chinch bugs, cye-spotted bud moth, false tarnished plant bug, lace bug, leafhoppers, leaf miners, lice, mealybugs, mites, pear psylla, scales, squash bug, thrips.

NOSEMA LOCUSTAE, GRASS-   Mfg: Reuter Labs
HOPPER SPORE

Grasshoppers, Mormon crickets.

OXAMYL, VYDATE   Mfg: DuPont

Aphids, black vine weevil, boll weevil, Colorado potato beetle, cottonleaf perforator, flea beetles, flea hopper, fungus fly, fungus gnats, green peach aphid, Japanese beetle, leafhoppers, leaf miners, mealybugs, mites, nematodes, pepper weevil, Royal palm bug, scales, thrips, whitefly.

OXYDEMETON-METHYL,   Mfg: Mobay
META-SYSTOX-R

Aphids, birch leaf miners, cone moth, corn rootworm, elm leaf beetle, flea beetles, flea hoppers, holly leaf miner, leafhopper, lygus bugs, mites, pine needle scale, Sitka spruce weevil, Texas citrus mites, thrips, whiteflies, white pine weevil.

OXYTHIOQUINOX, MORESTAN   Mfg: Mobay

Mitex, pear psylla, powdery mildew.

PARATHION   Mfg: Monsanto

Alfalfa caterpillar, alfalfa seed chalcid, alfalfa weevil, alfalfa weevil larvae, American cockroach, American plum borer, ants, aphids, apple flea weevil, apple lace bug, apple maggot, apple mealybug, apple red bugs, armyworms, artichoke plume moth, Asiatic garden beetle, avocado lace bug, avocado leafhopper, bagworms, bean leaf beetle, beet crown borer, black grass bug, black vine weevil, blister beetle, blossom anomala, blossom weevil, blueberry maggot, blueberry tipworm, boll weevil, bollworms, budworm, cabbage seedpod weevil, cankerworms, carrot rust fly, casebearer, catfacing insects, celery leaf tiers, cherry fruitworm, chinch bugs, citrus root weevil, climbing cutworms, clover head weevil, clover leaf weevils, codling moth, Colorado potato beetle, consperse stinkbug, corn earworms, corn rootworms, corn silkfly, cotton leaf perforators, cotton leaf worms, crane flies (larvae), cranberry fruitworm, cranberry tipworm, crickets, crown borer, cucumber beetle, cucumber beetle (banded), currant borer, cutworms, darkling ground beetles, diamondback moth, earwigs, European apple sawfly, European corn borer, European fruit lecanium scale, eye-spotted bud moth, false celery leaf tier, false chinch bug, fireworm, blackheaded flea beetles, fleahoppers, fruit flies, fruitworms, fuller rose beetle, grape berry moth, grape bud beetle, grasshoppers, green clover worm, green fruitworm, green June beetle, greenhouse leaf tier, Harlequin bug, hoplia beetles, hornworms, imported cabbageworm, Japanese beetle, Katydids, lacebugs, leaf beetles, leaf folders, leaf footed bugs, leafhoppers, leaf miners, leaf rollers, leaf tiers, lesser apple worm, lesser peach tree borer, lesser cornstalk borer, lima bean pod borer, little fire ant, loopers, mealybugs, melonworm, Mexican bean beetle, midges, mites, mosquitoes, obscure weevil, onion maggot, orange tortrix, orange dog, orangeworms, Oriental fruit moth, Pameras, pandemis moth, pea moth, pea weevil, pear psylla, peach bark beetle, peach tree borer, peach twig borer, pearbug, pecan leaf casebearer, pecan nut casebearer, pepper maggot, pickleworm, pink scavenger caterpillar, plant bugs, plum curculio, potato psyllid, potato tuberworm, psyllids, pumpkin bug, Quince curculio, raspberry crown borer, red-necked crane borer, red-necked peanutworm, rice leaf miner, rindworm, rose chafer, salt marsh caterpillar, San Jose scale, sawflies, Say's stinkbug, scales, seed corn maggot, shothold borer, snails (climbing), sorghum midge, sowbugs, spiders, spittlebugs, splitworm, spring tails, squash bugs, squash vine borer, stinkbugs, strawberry crown borer, strawberry rootworm, suckfly, Surinam cockroach, sweetclover weevil, symphylans, tadpole shrimp, tarnished plant bug, tent caterpillar, three-cornered alfalfa hopper, thrips, tip borers, tobacco budworm, tomato fruitworm, tomato pinworm, tortricid moths, tortrix moths, tussock moths, twig girdlers, vegetable weevil, velvet caterpillar, walnut caterpillar, walnut husk fly, webworms, weevils, white grubs, whiteflies, wireworms, wood weevil.

PERMETHRIN, AMBUSH, ATROBAN, ECTIBAN, POUNCE, PRAMEX　　　*Mfg: FMC and ICI*

Ants, apple aphids, armyworms, artichoke plume moth, asparagus beetle, bean leaf beetle, boll weevil, bollworms, budworms, cabbage looper, chinch bugs, cockroaches, codling moth, Colorado potato beetle, coneworms, corn earworms, corn borer, cotton aphids, cotton leaf perforator, crickets, crucifer weevil, cutworms, diamondback moth, European corn borer, eye gnats, filbertworm, flea beetles, fleahoppers, fowl mites, gnats, green cloverworm, green fruitworm, Heliothis Spp., hornworm, houseflies, imported cabbageworm, keds, leaf hoppers, leaf miners, leaf rollers, lice, loppers, lygus, mange mites, Mexican bean beetles, mosquitoes, mushroom flies, Nantucket pine tip moth, naval orangeworm, oblique banded leafroller, Oriental fruit moth, peach twig borer, pear psylla, periodical, circadas, pink bollworm, plum curculio, potato tuber worm, red banded leafroller, rose chafer, scabies, seed bugs, sheep keds, soybean looper, spiders, stable flies, stalk borers, stem weevils, tarnished plant bug, tentiform leafminer, termites, thrips, ticks, tomato fruitworm, velvetbean caterpillars, walnut husk fly, whiteflies.

PETROLEUM OIL, DORMANT AND SUMMER OILS   *Mfg: Numerous*

Aphids, apple red bug, black scales, brown apricot scale, brown mite, bud mite, citricola scale, codling moth, European fruit lecanium scales, European red mite, fig scale, frosted scale, fruit tree leaf roller, Italian pear scale, leaf roller, lichens, mealybugs, moss oleander scale, olive scale, Pacific mite, parlotoria, pear leaf blister mite, pear psylla, purple scale, Putnam scale, red scale, red spider mite, rust mite, San Jose scale, Scurfy scale, terrapin scale, twig borers, treehoppers, two-spotted mite, yellow scale, whiteflies.

PHENAMIPHOS, NEMACUR   *Mfg: Mobay*

Nematodes and thrips.

PHORATE, THIMET   *Mfg: American Cyanamid*

Alfalfa weevil, aphids, beet root maggot, black cutworm, clearwinged borers, Colorado potato beetle, corn leaf aphid, corn rootworm, cottonwood leaf beetle, cottonwood twig borer, European corn borer, grasshoppers, leaf beetle, Hessian fly, leafhoppers, leaf miners, lygus bugs, Mexican bean beetle, mites, Nantucket pine tip moth, pea leaf miners, planthoppers, psyllids, root lesion, nematodes, seed corn maggots, seed corn beetles, Southern corn rootworm, Southern potato wireworm, spotted alfalfa aphid, sugar beet root maggot, thrips, white grubs, whiteflies, wireworms.

PHOSALONE, ZOLONE   *Mfg: Rhone Poulenc*

Aphids, apple maggot, cherry fruit fly, codling moth, Colorado potato beetle, European corn borer, fruit tree leaf roller, hickory shuckworm, leafhoppers, mites, omnivorous leaf roller, Oriental fruit moth, leaf miners, orange dog, peach twig borer, pear psylla, pecan nut casebearer, pecan spittlebug, pecan weevil, pecan phylloxera, plume moth, plum curculio, red-banded leaf roller, thrips, whitefly.

PHOSMET, IMIDAN   Mfg: Stauffer

Alfalfa weevil, aphids, apple maggot, birch leaf miner, blueberry maggots, boll weevil, cattle grubs, cherry fruit fly, codling moth, Colorado potato beetle, corn rootworm, cranberry fruitworms, Eastern tent caterpillar, elm spanworm, European pine shoot moth, green fruitworm, grubs, gypsy moth, grape berry moth, hornflies, Japanese beetle, leafhoppers, leaf rollers, lice, navel orangeworms, oblique banded leaf rollers, orange tortrix, Oriental fruit moth, pea leaf weevil, pea weevil, peach twig borer, pear psylla, pecan nut casebearer, plum curculio, potato flea beetle, red humped caterpillars, scabies, scales, spring cankerworm, sweet potato weevil, tarnished plant bug, ticks.

PHOSPHAMIDON   Mfg: Chevron

Aphids, codling moth, Colorado potato beetle, European red mite, flea beetles, fleahoppers, fruit tree leaf roller, imported cabbageworm, leafhoppers, leaf miners, lygus bug, potato tuberworm, San Jose scale, spider mites, stinkbugs, thrips, walnut huskfly, whiteflies.

PIRIMIPHOS-METHYL, ACTELLIC   Mfg: ICI

Almond moth, angoremoris grain moth, cigarette beetle, confused flour beetle, corn sap beetle, flat grain beetle, granary weevil, hairy fungus beetle, Indian meal moth, maize weevil, merchant grain beetle, red flour beetle, rice weevil, sawtoothed grain beetle.

PROFENOSFOS, CURACRON   Mfg: CIBA-Geigy

Armyworm, boll weevil, bollworms, cabbage looper, cotton leaf perforator, mites, tobacco bud worms and whiteflies.

PROPARGITE, OMITE, COMITE   Mfg: Uniroyal

Mites.

PROPETAMPHOS, SAFROTIN   Mfg: Sandoz

Ants, cockroaches, crickets, earwigs, fireants, silverfish, spiders.

PROPOXUR, BAYGON   Mfg: Mobay

Ants, billbugs, booklice, cadelle, carpet beetles, centipedes, chinch bugs, cigarette beetle, clover mites, cockroaches, crickets, drugstore beetles, earwigs, fleas, fire brat, flies, flour moths, hornets, Indian meal moth, midges, millipedes, mole crickets, mosquitoes, saw-toothed grain beetle, scorpions, silverfish, sod webworms, sowbugs, spiders, ticks, wasps, webworms.

PYRAMDRON, AMDRO, MAX FORCE   *Mfg: American Cyanamid*

Imported fire ant, cockroaches.

PYRENONE (Combination of piperonyl butoxide and pyrethrins.)
                                      Mfg: Fairfield-American,
                                      MGK, Prentiss and
                                      others.

Ants, aphids, asparagus beetle, beetles, Blister beetles, boxelder bugs, cabbage looper, cabbageworm, Cadelles centipedes, cankerworms, cigarette beetles, Colorado potato beetle, crickets, cucumber beetle, diamondback moth, earwigs, firebrats, fireworms, flea beetles, fleas, flies, flour beetles, fruit flies, gnats, grain moths, gypsy moth, Harlequin bugs, horn flies, hornets, horseflies, leafhoppers, leaf tier, lice, med moths, medworms, Mexican bean beetle, mosquitoes, moths, plume moth, psoceds, roaches, silverfish, spiders, stable fleas, stinkbugs, stored grain insects, tarnished plant bug, vinegar flies, wasps, waterbugs, webworms, weevils, whiteflies.

PYRETHRIN   *Mfg: Fairfield-American,*
                                      MGK, Prentiss and
                                      others.

Ants, aphids, asparagus beetle, cabbage looper, cabbageworm, celery leaf tier, celery looper, cheese mites, Colorado potato beetle, crickets, cucumber beetle, diamondback moth, fireworm, flea beetles, flies, fruit flies, gnats, grain moths, Harlequin bug, Hawaiian beet webworm, leafhopper, leaf roller, lice, Mexican bean beetle, midge, mosquitoes, psyllid, roaches, silverfish, sod webworms, spiders, thrips, ticks, twelve-spotted beetle, wasps, webworms.

RESMETHRIN, SBP 1382   *Mfg: Penick Inc.*

Ants, aphids, centipedes, dog fleas, earwigs, fleas, flies, flour beetles, gnats, grain moths, hornets, Japanese beetle, leafhoppers, mosquitoes, moths, plant bugs, roaches, silverfish, sowbugs, spiders, spittlebugs, thrips, wasps, whiteflies.

ROTENONE   *Mfg: Penick and others*

Colorado potato beetle, fire ants.

SODIUM FLUOALUMINATE, KRYOCIDE

Mfg: Pennwalt

Armyworm, bean leaf beetle, bean leaf roller, blister beetle, cabbage looper, cabbageworm, codling moth, corn earworm, cranberry weevil, cucumber beetle, cutworms, darkling ground beetle, diamondback moth, earwigs, flea beetles, flea weevil, fruitworm, Fuller rose beetle, garden tortrix, grape leaf folder, gypsy moth, Holcocera, hornworm, katydid, leaf rollers, melonworm, Mexican bean beetle, orange dog, orange tortrix, pepper weevil, pickleworm, pinworm, plum curculio, rose weevil, strawberry root weevil, tobacco budworm, Western skeletonizer.

SULPROFOS, BOLSTAR

Mfg: Mobay

Armyworms, cotton bollworm, fleahoppers, green clover worm, Heliothis spp., leafhoppers, lygus, Mexican bean beetle, pink bollworms, plant bugs, three cornered alfalfa hopper, trips, tobacco budworm, velvetbean caterpillar.

TELONE II

Mfg: Dow

Bacterial canker, nematodes, quackgrass, soil insects, soil rot, verticillium wilt.

TEMEPHOS, ABATE

Mfg: American Cyanamid

Blackfly, buffalo gnat, mosquitoes, thrips, midges, gnats, punkies, sand flies.

TERBUFOS, COUNTER

Mfg: American Cyanamid

Billbugs, corn rootworms, nematodes, seed corn beetle, seed corn maggots, symphylans, white grubs, wireworms.

TETRACHLORVINPHOS, GARDONA, RABON

Mfg: Fermenta

Flies, lice.

TETRADIFON, TEDION

Mfg: Uniroyal

Mites.

THIODICARB, LARVIN

Mfg: Union Carbide

Armyworms, bean leaf beetle, boll weevil, cabbage looper, cotton leafhoppers, cotton leaf perforator, green clover worm, lygus, Mexican bean beetle, pink bollworms, podworms, soybean looper, stinkbugs, tobacco budworms.

TRALOMETRIN, SCOUT   Mfg: Hoechst-Roussel

Boll weevil, bollworm, cotton leaf perforator, pink bollworm, tobacco budworm.

TRICHLORFON, DIPTEREX, DYLOX,   Mfg: Mobay, Upjohn
PROXOL

Alfalfa plant bug, alfalfa webworm, armyworm, artichoke plume moth, bagworms, banana caterpillar, beet armyworm, beet webworm, birch leaf miner, black fleahopper, budworms, bulb fly, California orange dog, cattle grubs, chinch bug, citrus looper, climbing cutworms, cockroaches, cotton fleahopper, cotton leaf perforator, cotton leafworm, crickets, cutworms, darkling ground beetle, diamondback moth, dipterous leaf miner, elm leaf beetle, fleas, forest tent caterpillar, green June beetle, gypsy moth, horn fly, hornworm, horse bots, houseflies, imported cabbageworm, leafhoppers, leaf miners, leaf rollers, lice, lygus, Mexican bean beetle, Nantucket pine tip moth, Narcissus bulb fly, orange tortrix, pepper maggot, salt marsh caterpillar, sarwigs, sod webworms, squash bugs, stinkbug, tarnished plant bug, tent caterpillars, thrips, ticks, tobacco budworm, tomato hornworm, varigated cutworm, webworms, Western bean cutworm, Western yellow striped armyworm, Zimmerman pine tip moth.

TRIMETHACARB, BROOT   Mfg: Union Carbide

Corn rootworm

VIKANE   Mfg: Dow

Bedbugs, carpet beetles, clothes moths, cockroaches, deathwatch beetles, mice, old house borers, powder-post beetles, termites.

VORLEX   Mfg: Nor-Am

Fungi, nematodes, soil insects, symphylans, weeds, wireworms.

NOTES

HERBICIDES

HERBICIDES

1. **ABELIA**
   Betasan
   Dacthal
   Devrinol
   ProGrow
   Ronstar
   Surflan

2. **ABUTILON**
   Ronstar

3. **ACACIA**
   Treflan

4. **ACER**
   Devrinol

5. **AGAPANTHUS**
   Devrinol
   Ronstar

6. **AGERATUM**
   CIPC
   Dacthal
   Eptam
   Devrinol
   Eptam
   Treflan

7. **AJUGA**
   Betasan
   Devrinol
   Eptam
   Fusilade
   Poast
   Ronstar

8. **ALDER**
   Fusilade

9. **ALMOND**
   (Flowering)
   Casoron
   Devrinol
   Enide
   Fusilade

10. **ALPINE CUR-**
    RENT
    Poast

11. **ALYSSUM**
    Betasan
    CIPC
    Dacthal
    Enide
    Eptam
    Fusilade
    Poast
    Treflan

12. **AMARANTHUS**
    Eptam

13. **AMERICAN**
    BITTERSWEET
    Poast

14. **AMUR**
    CORKTREE
    Casoron

15. **APPLE**
    (Flowering)
    CIPC
    Devrinol
    Enide
    Surflan
    Treflan

16. **ARALIA**
    Ronstar

17. **ARBOR VITAE**
    (Thuja)
    Amiben
    Amino Triazole
    Amizine
    Casoron
    CIPC
    Dacthal
    Devrinol
    Dual
    Enide
    Fusilade
    Goal
    Pennant
    Poast
    ProGrow
    Ronstar
    Roundup
    Rout
    Simazine
    Surflan
    Treflan

18. **ARCTOSTA-**
    PHYLOS
    Ronstar

19. **ARCTOTIS**
    (African Daisy)
    Treflan

20. **ARROW WOOD**
    Poast

21. **ASH**
    Amiben
    Amizine
    Casoron

Dacthal
Devrinol
Enide
Fusilade
Karmex
Poast
Ronstar
Treflan

22. **ASPARAGUS**
Devrinol

23. **ASTER**
Betasan
CIPC
Dacthal
Devrinol
Enide
Eptam
Treflan

24. **ASTILBE**
CIPC

25. **AUCUBA**
Devrinol
Dual
Fusilade
Pennant
XL

26. **AZALEA**
Amiben
Betasan
Casoron
CIPC
Dacthal
Devrinol
Dual
Enide
Eptam
Fusilade
Kerb
Pennant
Poast
Ronstar

Roundup
Surflan
Treflan

27. **AZARA**
Betasan

28. **BABYSBREATH**
Dacthal
Enide

29. **BABY TEARS**
Enide

30. **BALSAM**
Amizine
CIPC
Eptam
Treflan

31. **BAMBOO**
(Nandina)
Pennant
Ronstar
Surflan

32. **BARBERRY**
(Berberis)
Casoron
CIPC
Dacthal
Dual
Enide
Eptam
Fusilade
Pennant
Poast
Ronstar
Rout
Simazine
Surflan
Treflan

33. **BAYAN**
Ronstar

34. **BEAR BERRY**
Fusilade

35. **BEAUTY BRUSH**
Casoron
Enide

36. **BEECH**
Amizine
Enide
Fusilade

37. **BEEFSTEAK
PLANT**
Ronstar

38. **BEGONIA**
CIPC
Enide
Eptam
Poast
Surflan

39. **BIG BLUE
LILYTURF**
Surflan

40. **BIRCH**
Casoron
CIPC
Dacthal
Devrinol
Enide
Fusilade
Poast
Ronstar
Treflan

41. **BIRD OF
PARADISE**
Devrinol
Surflan

42. **BLEEDING HEART**
Dacthal
Poast

43. **BLOOD LEAF**
Dacthal

44. **BLUEBELLS**
CIPC

45. **BOTTLEBRUSH**
Devrinol
Ronstar
Simazine
Surflan
XL

46. **BOUGAIN-VILLEA**
Devrinol
Ronstar
Rout

47. **BOX ELDER**
Casoron
Simazine

48. **BOX WOOD (Buxus)**
Betasan
CIPC
Casoron
Dacthal
Devrinol
Dual
Eptam
Fusilade
Pennant
Poast
Prograss
Ronstar
Roundup
Rout

Surflan
Treflan

49. **BRASSAIA**
Ronstar
XL

50. **BRUSH CHERRY**
Surflan

51. **BUCKTHORNE**
Poast

52. **BUGLOSS**
Dacthal

53. **CACTUS**
Poast
Surflan

54. **CALENDULA**
Betasan
Fusilade
Treflan

55. **CAMELLIA**
Casoron
CIPC
Dacthal
Devrinol
Enide
Eptam
Pennant
Poast
ProGrow
Ronstar
Treflan

56. **CAMPANULA (Bell Flower)**
Betasan
Dacthal
Poast

57. **CANDY TUFT**
Betasan
Dacthal
Rout

58. **CANNA**
Poast

59. **CAPEWEED**
Surflan
XL

60. **CARAGANA (Pea Tree)**
Casoron
Karmex
Simazine

61. **CAROB**
Simazine

62. **CARNATIONS**
CIPC
Eptam
Poast
Ronstar
Treflan

63. **CARPOBROTUS**
Devrinol
Ronstar

64. **CATALPA**
Poast

65. **CEANOTIIUS**
Enide
Ronstar
Surflan

66. **CEDAR**
Casoron
CIPC
Devrinol

Enide
Goal
Karmex
Poast
Ronstar
Simazine
Surflan
Treflan

67. **CELOSIA
    (Cockscomb)**
    Amiben
    Poast

68. **CEMTAUREA-
    BACHELOR'S
    BUTTON, DUSTY
    MILLER**
    Betasan
    Treflan

69. **CHAMAECY-
    PARIS**
    Eptam
    ProGrow

70. **CHERRY
    (Flowering)**
    CIPC
    Devrinol
    Enide
    Fusilade
    Poast
    Surflan

71. **CHERRY-
    LAUREL**
    Dymid
    Enide
    Treflan

72. **CHESTNUT**
    Dacthal
    Enide
    Treflan

73. **CHOKEBERRY**
    Poast

74. **CHRYSANTHE
    MUMS**
    Amiben
    CIPC
    Dacthal
    Devrinol
    Enide
    Eptam
    Poast
    Surflan
    Treflan

75. **CHINESE ELM**
    Amizine

76. **CINQUEFOIL
    (Potentilla)**
    Dacthal
    Fusilade
    Pennant
    Poast
    ProGrow
    Treflan

77. **CLEYERA
    JAPONICA**
    Casoron

78. **COLEUS**
    Casoron
    CIPC
    Dacthal
    Poast

79. **COLUMBINE**
    CIPC
    Dacthal

80. **CORALBELLS**
    Betasan
    CIPC
    Dacthal
    Poast

81. **COREOPSIS-
    CALLIOPIS-TICK
    SEED**
    Dacthal
    Treflan

82. **CORNUS**
    ProGrow

83. **COROMANDEL**
    Ronstar

84. **COSMOS**
    Dacthal
    Treflan

85. **COTONEASTER**
    Amiben
    Casoron
    CIPC
    Dacthal
    Devrinol
    Dual
    Enide
    Fusilade
    Kerb
    Pennant
    Poast
    ProGrow
    Ronstar
    Rout
    Simazine
    Surflan
    Treflan

86. **COTTONWOOD**
    Casoron
    Dacthal
    Enide
    Karmex
    Treflan

87. **COYOTE BRUSH**
    Surflan

88. **CRAB-APPLE**
   (Malus)
   (Flowering)
   Casoron
   CIPC
   Dacthal
   Devrinol
   Dual
   Enide
   Fusilade
   Pennant
   Poast
   Ronstar
   Roundup

89. **CRANBERRY**
   BUSH
   Poast

90. **CRAPE MYRTLE**
   (Lagerstromis)
   Devrinol
   Fusilade
   Pennant
   Poast
   Ronstar
   Rout

91. **CRATAEGUS**
   Devrinol

92. **CREAM CARPET**
   Enide

93. **CROTON**
   Rout

94. **CROWNVETCH**
   Poast

95. **CRYPTOMERIA**
   Surflan

96. **CUPHEA**
   Dacthal

97. **CURRENT,**
   ALPINE
   Poast

98. **CYPRESS**
   Betasan
   Casoron
   Dacthal
   Devrinol
   Enide
   Fusilade
   Ronstar
   Surflan
   Treflan

99. **DAFFODIL**
   Betasan
   CIPC
   Sinox

100. **DAHLIA**
   Amiben
   Betasan
   CIPC
   Dacthal
   Devrinol
   Enide
   Eptam
   Poast
   Treflan

101. **DAISY**
   Betasan
   Devrinol
   Enide
   Ronstar
   Surflan
   Treflan
   XL

102. **DAPHNE**
   Betasan

103. **DAYLILIES**
   Amiben
   Eptam

104. **DELOSPERMA**
   Devrinol

105. **DELPHINIUM**
   (Larkspur)
   CIPC
   Dacthal
   Enide

106. **DEUTZIA**
   Casoron
   Dacthal
   Treflan

107. **DICHRONDA**
   Betasan
   Devrinol
   Enide

108. **DIEFFENBACHIA**
   Ronstar

109. **DIMORTHECA**
   Treflan

110. **DIOSMA**
   Rout

111. **DODONAEA**
   Ronstar

112. **DOGWOOD**
   Amiben
   Amizine
   Casoron
   CIPC
   Dacthal

Devrinol
Dual
Enide
Eptam
Fusilade
Pennant
Poast
Ronstar
Simazine
Treflan

113. ELAEAGNUS
Poast

114. ELM
Amizine
Casoron
Dacthal
Fusilade
Karmex
Ronstar
Simazine

115. ENKIANTHUS
CIPC

116. ERICA
(Heath)
ProGrow

117. ERYSIMUM
Devrinol

118. ESCALLONIA
Pennant
Rout
Surflan

119. EUCALYPTUS-GUM
Dacthal
Devrinol
Enide
Fusilade

Poast
Ronstar
Simazine
Surflan
Treflan

120. EUONYMUS
Amiben
Casoron
CIPC
Dacthal
Devrinol
Dual
Enide
Eptam
Fusilade
Kerb
Pennant
Poast
ProGrow
Ronstar
Roundup
Rout
Surflan
Treflan

121. EUPHORBIA
Rout

122. FATSHEDERA
Surflan

123. FEIJOA-PINEAPPLE GUAVA
Treflan

124. FERNS
Karmex
Poast

125. FEVERFEW
Dacthal
Enide

126. FICUS
Devrinol
Rout

127. FIR
Amino Triazole
Amizine
Asulox
Atrazine
CIPC
Dacthal
Devrinol
Eptam
Fusilade
Goal
Karmex
Kerb
Modown
Pennant
Poast
Ronstar
Roundup
Simazine
Surflan
Treflan

128. FORGET-ME-NOT
CIPC
Dacthal
Treflan

129. FORSYTHIA-GOLDEN BELL
Casoron
CIPC
Dacthal
Devrinol
Dual
Enide
Fusilade
Kerb
Pennant
Poast
ProGrow
Ronstar

Surflan
Treflan

130. **FOUR-O-CLOCK**
Dacthal
Treflan

131. **FOXGLOVE**
Dacthal
Enide

132. **FREESIA**
Betasan

133. **FUCHSIA**
CIPC

134. **GAILLARDIA**
Dacthal
Treflan

135. **GARDENIA
(CAPE JASMINE)**
Casoron
Devrinol
Fusilade
Poast
ProGrow
Ronstar
Rout
Surflan

136. **GANZANIA**
Betasan
Devrinol
Eptam
Poast
Ronstar
Surflan
XL

137. **GELSEMIUM**
Rout

138. **GERANIUM**
CIPC
Dacthal
Devrinol
Enide
Fusilade
Poast
Surflan

139. **GINKGO**
Ronstar
Surflan

140. **GLADIOLUS**
Betasan
CIPC
Dacthal
Devrinol
Karmex
Poast
TCA
Treflan

141. **GLEDITSIA
(Honey Locust)**
Devrinol

142. **GOLD DUST**
Dacthal

143. **GOLDEN TUFT**
Dacthal

144. **GRABERI**
Surflan

145. **HACKBERRY**
Casoron
Ronstar

146. **HAWTHORN**
Dacthal
Devrinol
Ronstar

Rout
Surflan
Treflan

147. **HEATH**
Ronstar

148. **HEATHER**
Casoron
CIPC
Dacthal
Devrinol
Enide
Ronstar

149. **HEDERA**
Devrinol

150. **HELIOTROPE**
Enide

151. **HEMLOCK**
Amiben
Amizine
CIPC
Devrinol
Enide
Eptam
Fusilade
Goal
Kerb
Modown
Poast
Ronstar
Simazine
Treflan

152. **HERNIARIA**
Poast

153. **HIBISCUS-
ROSE MALLOW**
Devrinol
Enide

Rout
Surflan

154. HICKORY
Amiben

155. HOLLY (Ilex)
Amiben
Amizine
Betasan
Casoron
CIPC
Dacthal
Devrinol
Dual
Enide
Eptam
Fusilade
Kerb
Pennant
Poast
ProGrow
Ronstar
Roundup
Rout
Simazine
Surflan
Treflan

156. HONEYSUCKLE
Casoron
CIPC
Dacthal
Devrinol
Dual
Karmex
Pennant
Poast
Ronstar
Simazine
Surflan

157. HOP-BUSH
Enide

Simazine
Surflan
Treflan

158. HORSE CHEST-NUT
Fusilade

159. HOSTA
Devrinol
Fusilade
Poast

160. HYDRANGEA
CIPC
Dacthal
Enide
Fusilade
ProGrow

161. HYPERICUM (St. Johnswort)
Betasan
Devrinol
Enide
Eptam
Pennant
Rout
Surflan
XL

162. ICE PLANT
Betasan
Devrinol
Enide
Eptam
Poast
Surflan
Treflan

163. ILIMA
Ronstar
XL

164. IMPATIENS
CIPC
Fusilade
Poast
Surflan

165. IRIS
Amiben
CIPC
Dacthal
Fusilade
Karmex
Poast

166. IVY
Amiben
Betasan
Casoron
CIPC
Dacthal
Devrinol
Dual
Enide
Eptam
Fusilade
Pennant
Poast
Ronstar
Surflan
XL

167. IXORA
Treflan

168. JACK-IN-THE PULPIT
Poast

169. JAPANESE LARCH
Devrinol

170. JASMINE
Betasan
Devrinol

Enide
Pennant
Ronstar
Rout
Surflan

171. **JOJOBA**
Poast

172. **JUNIPERS**
Amiben
Amino Triazole
Amizine
Asulox
Betasan
Casoron
CIPC
Dacthal
Devrinol
Dual
Enide
Eptam
Fusilade
Goal
Kerb
Lasso
Pennant
Poast
ProGrow
Ronstar
Rout
Simazine
Surflan
Treflan
XL

173. **JUSTICA**
Ronstar

174. **KINNIKINNICK-
CORNUS
AMOMUM**
Casoron
Rout

175. **LABURNUM**
Devrinol

176. **LANTANA**
Dacthal
Devrinol
Pennant
Rout

177. **LAPALAPA**
Ronstar

178. **LAUREL-
SWEET BAY**
Casoron
CIPC
Dual
Enide
Surflan
Treflan

179. **LAURUSTINUS**
Surflan

180. **LAVENDER
COTTON**
Dacthal

181. **LEUCOTHOE**
Casoron
CIPC
Devrinol
Dual
Eptam
Pennant
Ronstar
Surflan

182. **LILAC
(Syringa)**
Casoron
CIPC
Dacthal
Dual
Enide

Eptam
Pennant
Poast
Ronstar
Roundup
Surflan
Treflan

183. **LILIES**
Amiben
CIPC
Dacthal
Eptam
Pennant

184. **LILY-OF-
THE-NILE**
Ronstar
Surflan
XL

185. **LILY OF
THE VALLEY**
Poast

186. **LILYTURF**
Fusilade

187. **LINDEN-
BASSWOOD**
Casoron
Eptam
Poast

188. **LIRIOPE**
Devrinol
Dual
Fusilade
Pennant
Poast

189. **LOBELIA**
Enide
Poast
Treflan

190. **LOCUST**
Casoron
CIPC
Dacthal
Enide
Fusilade
Poast
Simazine
Treflan

191. **LONDON-
PLANE TREE**
Amizine
Treflan

192. **LOOSESTRIFE**
Enide

193. **LUPINE**
Dacthal
Lupine

194. **MAGNOLIA**
Amiben
Casoron
CIPC
Dacthal
Eptam
Fusilade
Poast
Ronstar
Roundup
Surflan

195. **MALUS**
Devrinol

196. **MANZANITA**
Surflan

197. **MAPLES
(ACER)**
Amizine
Casoron
CIPC
Dacthal

Devrinol
Dual
Enide
Eptam
Fusilade
Pennant
Poast
Ronstar
Roundup
Surflan
Treflan

198. **MARIGOLD**
Amiben
Betasan
CIPC
Dacthal
Enide
Eptam
Fusilade
Poast
Surflan
Treflan

199. **METALLIC
PLANT**
Ronstar

200. **MICROBIOTA**
Rout

201. **MONDO-GRASS**
Poast

202. **MONEY TREE**
Ronstar

203. **MONEYWORT**
Poast

204. **MORNING
BRIDE-SCAB-
IOSA**
Dacthal
Treflan

205. **MORNING-
GLORY**
Dacthal
Treflan

206. **MOSS SAND-
WORT**
Poast

207. **MOTHER OF
THYME**
Dacthal

208. **MOUNTAIN ASH**
Fusilade

209. **MOUNTAIN
LAUREL
(Kalmic)**
Dacthal
Dual
Pennant
Surflan

210. **MULTIFLORA
ROSE**
Amiben

211. **MUSSAENDA**
Ronstar

212. **Myrtle
(Vinca)**
Betasan
CIPC
Devrinol
Enide
Ronstar
Surflan

213. **NANDINA**
Casoron
Devrinol
Poast
Surflan

214. **NANNYBERRY**
Poast

215. **NARCISSUS**
Betasan
CIPC
Devrinol
Karmex

216. **NASTURTIUM**
Dacthal
Eptam
Treflan

217. **NICOTIANA**
Treflan

218. **NINEBARK**
Pennant
Poast

219. **NURSERY STOCK (All kinds)**
CIPC
Kerb

220. **OAK**
Amiben
Casoron
Dacthal
Devrinol
Dual
Enide
Eptam
Fusilade
Kerb
Pennant
Poast
Prograss
Ronstar
Roundup
Simazine
Surflan
Treflan

221. **O'CONNOR'S LEGUME**
Fusilade

222. **OHAI**
Ronstar

223. **OLEANDER**
Amino Triazole
Amizine
Devrinol
Enide
Paraquat
Poast
ProGrow
Ronstar
Rout
Simazine
Surflan
XL

224. **OLIVE**
Devrinol
Enide
XL

225. **OPHIOPOGON**
Ronstar

226. **OREGON GRAPE (Mahonia)**
CIPC
Devrinol
Enide
Fusilade
Ronstar
Simazine
Surflan

227. **ORNAMENTAL PEPPER**
Poast

228. **ORNAMENTALS (All kinds)**
Chloropicirin
Fusilade
Methyl Bromide
Vapam
Vorlex

229. **ORTHOSIPHON**
Ronstar

230. **OSAGE ORANGE**
Poast

231. **OSMANTHUS (Holly-Olive)**
Casoron
Devrinol
Dual
Pennant
Ronstar
Surflan

232. **OSTEOSPERUM**
Devrinol
Ronstar

233. **PACHISTIMA**
Casoron
Dacthal

234. **PACHYSANDRA-MOUNTAIN SPURGE**
Betasan
CIPC
Dacthal
Devrinol
Dual
Eptam
Fusilade
Pennant
Poast
Ronstar

235. **PALM**
Devrinol
Ronstar
Simazine

236. **PANSY**
Betasan
CIPC
Eptam
Poast
Surflan

237. **PAPERBARK TREE**
Ronstar

238. **PEACH (Flowering)**
Devrinol
Enide
Fusilade

239. **PEAR (Flowering)**
Devrinol
Fusilade
Poast
Surflan

240. **PEONIES**
Amiben
CIPC
Dacthal
Enide

241. **PEPPERMINT**
Surflan

242. **PERIWINKLE-VINCA**
Betasan
Enide
Eptam
Fusilade

Poast
Ronstar
Surflan
Treflan

243. **PETUNIA**
CIPC
Dacthal
Devrinol
Enide
Eptam
Fusilade
Poast
Surflan
Treflan

244. **PHILODENDRON**
Enide

245. **PHLOX**
Enide
Treflan

246. **PHOTINIA**
Casoron
Devrinol
Fusilade
Pennant
Poast
Ronstar
Rout
Surflan
XL

247. **PIERIS (Andromeda)**
Amiben
CIPC
Dacthal
Dual
Eptam
Pennant
ProGrow
Ronstar

Simazine
Surflan
Treflan

248. **PINES**
Amiben
Amino Triazole
Amizine
Asulox
Atrazine
Betasan
Casoron
CIPC
Devrinol
Enide
Eptam
Fusilade
Goal
Karmex
Kerb
Modown
Oust
Pennant
Poast
ProGrow
Ronstar
Roundup
Rout
Simazine
Surflan
Treflan

249. **PITTOSPORUM (Mock Orange)**
Casoron
CIPC
Dacthal
Devrinol
Enide
Fusilade
Poast
ProGrow
Ronstar
Rout

Surflan
Treflan

250. **PLUM**
(Flowering)
CIPC
Devrinol
Enide
Fusilade

251. **PODOCARPUS**
Asulox
Dacthal
Devrinol
Eptam
Fusilade
Pennant
Poast
Ronstar
Rout
Surflan

252. **POINCIANA**
Ronstar

253. **POKER PLANT**
Dacthal

254. **PONDS AND**
LAKES
Acrolein
Aquazine
Banvel
Casoron
Copper Compounds
Copper Sulfate
Curtrine
2,4-D
Dalapon
Dichlone
Diquat
Fenac
Hydrothol-
Aquathol

Karmex
Monuron
Petroleum solvents
Rodeo
Roundup
Simazine
Sonar
Spike
Velpar
Xylene

255. **POPLAR**
Casoron
CIPC
Dacthal
Devrinol
Poast
Ronstar

256. **POPPY**
Treflan

257. **PORTULACA**
Poast

258. **PREMISES**
Ametryne
Amino Triazole
Ammate
Arsenal
Asulox
Atrazine
Banvel
Borates
Bromoxynil
Cacodylic Acid
Casoron
Crossbow
Dalapon
Diquat
DSMA
2,4-D
2,4-DP
Endothal

Fenac
Fusilade
Garlon
Hyvar
Igran
Karmex
Krenite
Krovar
Lorox
MCPA
MSMA
Oust
Paraquat
Poast
Pramitol
Prometryne
Rodeo
Roundup
Simazine
Sodium Borates
Sodium Chlorate
Spike
Stomp
TCA
Telar
Telok
Tordon
Urox
Velpar

259. **PRIMROSE**
Betasan
Dacthal

260. **PRIVET-**
LIGUSTRUM
Betasan
Casoron
CIPC
Dacthal
Devrinol
Dual
Enide
Fusilade

63

Pennant
Poast
ProGrow
Ronstar
Roundup
Rout
Surflan
Treflan

261. **PROTEA**
Ronstar

262. **PRUNUS**
Devrinol

263. **PURPLE-LEAF WINTER CREEPER**
CIPC
Surflan

264. **PYRACANTHA (Firethorne)**
Amino Triazole
Amizine
Betasan
Casoron
Devrinol
Dual
Enide
Fusilade
Pennant
ProGrow
Ronstar
Surflan
Treflan
XL

265. **QUERCUS**
Ronstar

266. **QUINCE**
Casoron
Fusilade

267. **RANNUCULUS-BUTTERCUP**
Betasan

268. **RAPHIOLEPIS**
Devrinol
XL

269. **RED BUD**
Dacthal
Enide
Fusilade
Treflan

270. **RED ROOT**
Surflan

271. **REDWOOD**
Modown
Simazine
Surflan

272. **RHODODEN-DRON**
Amiben
Casoron
CIPC
Dacthal
Devrinol
Dual
Enide
Eptam
Fusilade
Kerb
Pennant
Poast
ProGrow
Ronstar
Rout
Surflan
Treflan

273. **ROCKROSE**
Casoron

274. **ROSEBAY**
Dual
Pennant

275. **ROSES**
Casoron
Dacthal
Devrinol
Dual
Enide
Fusilade
Pennant
Ronstar
Surflan
Treflan

276. **RUDBECKIA-CONE FLOWER**
Treflan

277. **RUSSIAN OLIVE**
Casoron
Dacthal
Enide
Karmex
Poast
Ronstar
Simazine
Treflan

278. **SAKAKI**
Treflan

279. **SALVIA**
CIPC
Dacthal
Enide
Poast
Treflan

280. **SANDANKWA**
Betasan

281. **SANDCHERRY**
Pennant
Poast

282. **SCARLET SAGE**
Dacthal

283. **SCHEFFLERA**
Rout

284. **SEDUM**
Betasan
Dacthal
Devrinol
Eptam
Fusilade
Poast
Ronstar
Surflan

285. **SENSITIVE PLANT**
Poast

286. **SERVICE BERRY**
Poast

287. **SHRIMP PLANT**
Surflan

288. **SNAPDRAGONS**
Amiben
CIPC
Dacthal
Enide
Poast
Treflan

289. **SNOW-IN-SUMMER**
Poast

290. **SNOW ON THE MOUNTAIN**
Treflan

291. **SASSAFRAS**
Amiben

292. **SPIDERWORT**
Dacthal

293. **SPINDLE TREE**
Poast

294. **SPIREA**
Amiben
Casoron
CIPC
Dacthal
Dual
Enide
Fusilade
Pennant
Poast
ProGrow
Treflan

295. **SPRENGER**
Poast

296. **SPRUCE (Picea)**
Amino Triazole
Amizine
Atrazine
Casoron
CIPC
Dacthal
Devrinol
Eptam
Fusilade
Goal
Modown
Pennant
Poast
Ronstar
Roundup
Rout
Simazine

Surflan
Treflan

297. **SQUAW CARPET**
Casoron

298. **STAR JASMINE**
Devrinol
Ronstar
Surflan

299. **STOCKS**
Betasan
Enide
Surflan
Treflan

300. **STONE CROP**
Dacthal
Surflan

301. **STRAWBERRY (Ornamental)**
Betasan
Enide
Eptam

302. **STRAWBERRY TREE**
Poast

303. **STRAWFLOWER**
Dacthal

304. **SUNDROPS**
Dacthal

305. **SUNFLOWER**
Dacthal
Treflan

306. **SWEET BROOM**
Pennant

307. **SWEET GUM**
Fusilade
Poast
Surflan

308. **SWEET OLIVE**
Ronstar

309. **SWEET PEA**
Betasan
Dacthal
Treflan

310. **SWEET WILLIAM**
Enide
Poast
Treflan

311. **SYCAMORE**
Amizine
CIPC
Dacthal
Enide
Poast
Treflan

312. **SYRINGA**
Fusilade

313. **SYZYGIUM**
Ronstar

314. **TAMARACH-JAPANESE LARCH**
Devrinol
Poast
Surflan
Treflan

315. **TARO VINE**
Ronstar

316. **TEA TREE**
Poast

317. **TIARE**
Ronstar

318. **TOBIRA**
Betasan
Treflan

319. **TRUMPET VINE**
Surflan

320. **TSUGA**
Ronstar

321. **TULIPS**
Betasan
CIPC
Karmex

322. **TULIP TREE**
Dacthal
Enide
Fusilade
Treflan

323. **TURF (General)**

Amino Triazole
Asulox
Atrazine
Balan
Banvel
Basagran
Betasan
Bromoxynil
Buctril
Cacodylic Acid
Calar
2,4-D
2,4-DP
Dacthal

Devrinol
DSMA
Endothal
Enide
Garlon
Kerb
MCPA
MCPP
MSMA
Poast
Po-San
Prograss
Prowl
Ronstar
Roundup
Sencor
Simazine
Surflan
Team
Tupersan
Turf-Cal
Turflon
Vapam
Vorlex
XL

324. **VERBENA**
Dacthal
Enide
Poast
Rout

325. **VIBURNUM**
Amiben
Amino Triazole
Amizine
CIPC
Dacthal
Devrinol
Dual
Enide
Eptam
Fusilade
Pennant

66

Poast
ProGrow
Ronstar
Rout
Surflan
Treflan

326. **VIOLETS**
Dacthal

327. **WALL FLOWER**
Betasan

328. **WALNUT
(Black)**

Dacthal
Devrinol
Enide
Poast
Simazine
Treflan

329. **WANDERING
JEW**
Poast

330. **WAYFARING
TREE**
Poast

331. **WEDELIA**
Ronstar

332. **WEIGELA**
Casoron
Dacthal
Dual
Enide
Pennant
Rout
Surflan
Treflan

333. **WILLOW**
Casoron
Dacthal
Enide
Fusilade
Poast
Treflan

334. **WINTER
CREEPER**
CIPC
Surflan

335. **WORMWOOD**
Dacthal

336. **XYLOSMA**
Betasan
Devrinol
Ronstar
Surflan

337. **YARROW**
Dacthal

338. **YAUPON
(Ilex)**
Pennant

339. **YEW (Taxus)**
Amiben
Amino Triazole
Amizine
Asulox
Casoron
CIPC
Dacthal
Devrinol
Dual
Enide
Eptam
Fusilade
Goal
Kerb

Lasso
Pennant
Poast
ProGrow
Ronstar
Roundup
Simazine
Surflan
Treflan

340. **YUCCA**
Surflan

341. **ZINNIA**
Amiben
Betasan
CIPC
Dacthal
Devrinol
Enide
Eptam
Fusilade
Poast
Surflan
Treflan

NOTES

HERBICIDES

ACIFLUORFEN-SODIUM, BLAZER, TACKLE

Mfg: Rohm and Haas
Rhone Poulenc

Amaranth, balloonvine, beggarweed, buffalobur, Canadian thistle, carpetweed, citron, cocklebur, coffee weed, copperleaf, croton, fall pancium, field bindweed, Florida pursley, foxtails, galinsoga, gerkins, giant ragweed, groundcherry, hairy indigo, hemp sesbania, hophornbean, ironweed, jimsonweed, ladys thumb, lambsquarters, milkweed, morningglory, mustard, nightshade, pigweed, prostrate spurge, purple moon flower, purslane, ragweed, seedling johnsongrass, showy crotalaria, smell melon, smartweed, smooth pigweed, spring cucumber, Texas gourd, trumpet creeper, Virginia copperleaf, wild mustard, wild poinsettia.

ALACHLOR, LASSO

Mfg: Monsanto

Barnyard grass, black nightshade, carpetweed, crabgrass, fall panicum, Florida pursley, foxtails, goosegrass, hairy nightshade, pigweed, purslane, red rice, seedling signal grass, witchgrass, yellow nutgrass.

AMETRYNE, EVIK

Mfg: CIBA-Geigy

Annual broadleaves, annual grasses, Bachiaria, barnyard grass, cocklebur, crabgrass, Crotalaria, dallisgrass, fireweed, Flora's paintbrush, Florida pursley, foxtails, goosegrass, Japanese tea, jungle rice, kukaipuaa, lambsquarters, morningglory, mustard, nutsedge, panicum, pigweed, purslane, ragweed, rattlepod, shattercane, signal grass, sow thistle, Spanish needles, velvetleaf, wild pea bean, wild proso millet, wiregrass.

AMINO TRIAZOLE, AMITROLE

Mfg: Union Carbide and
American Cyanamid

Annual bluegrass, ash, barnyard grass, bermudagrass, big leaf maple, blackberry, bluegrasses, Canada thistle, carpetweed, catchfly, cattails, cheatgrass, chickweed, chrysanthemum weed, cocklebur, crabgrass, dandelion, dock, downybrome, fanweed, foxtail, goosegrass, groundsel, hempnettle, honeysuckle, horse nettle, horsetail rush, jimsonweed, knotweed, kochia, kudzuvine, lambsquarters, leafy spurge, locust, milkweed, mustard, nightshade, nutgrass, pharagmites, pigweed, plaintain, poison ivy, poison oak, puncturevine, purslane, quackgrass, ragweed, ryegrass, Salmon berry, shepherdspurse, smartweed, sow thistle, stinkweed, sumac, sunflower, tarweed, velvetgrass, volunteer alfalfa, water hyacinth, white cockle, white grass, white top, wild barley, wild buckwheat, wild cherry, wild oats, yellow rocket.

AMMONIUM SULFAMATE, AMMATE   Mfg: DuPont

Alder, ash, birch, bitter dock, blueweed, broomweed, brush control, cedar, cocklebur, crabgrass, elm, goldenrod, gum, hickory, jimsonweed, lambsquarters, larkspur, leafy spurge, maple, milkweed, oaks, pecan, perennial ragweed, pine, poison ivy, poison sumac, prickly lettuce, ragweed, shepherdspurse, sweetgum, willow and most annual weeds.

ASULAM, ASULOX   Mfg: Rhone Poulenc

Alexander grass, barnyard grass, broadleaf panicum, bullgrass, California grass, crabgrass, foxtails, goosegrass, guinea grass, horseweed, itchgrass, johnsongrass, panicum, paragrass, Raoulgrass, sandbur, Western bracken.

ATRAZINE, AATREX   Mfg: CIBA-Geigy and others

Ageratum, amaranth, barnyard grass, blackeyed susans, bluegrass, broomsedge, broomweed, burdock, Canada thistle, carpetweed, cheatgrass, chickweed, cinquefoil, cocklebur, crabgrass, cranesbell, dock, dog fennel, dogbane, downybrome, fanweed, fiddleneck, fireweed, fleabane, Flora's paintbrush, foxtail, galinsoga, goosegrass, groundcherry, groundsel, hoary cress, horseweed, jimsonweed, jungle rice, knotweed, kochia, ladys thumb, lambsquarters, little barley, little bluestern, medusahead, morningglory, mustard, nightshade, nutgrass, orchardgrass, paplo, pellitory weed, pigweed plaintain, pineapple weed, poverty weed, puncturevine, purple top, purslane, quackgrass, ragweed, rattlepod, redtop, Russian thistle, sagewort, sandbur, shepherdspurse, sicklepod, smartweed, smooth brome, sneezeweed, sow thistle, Spanish needles, spurge, sunflower, tall buttercup, tansy mustard, tumble mustard, velvetleaf, volunteer grains, watergrass, wild carrot, wild lettuce, wild oats, wiregrass, wirestem, witchgrass, Yarrow.

BARBAN, CARBYNE   Mfg: Velsicol

Annual ryegrass, canary grass, wild oats.

BENEFIN, BALAN   Mfg: Elanco

Annual bluegrass, barnyard grass, carelessweed, carpetweed, chickweed, crabgrass, crowfoot grass, deadnettle, Florida pursley, foxtail, goosegrass, johnsongrass (from seed), jungle rice, knotweed, lambsquarters, pigweed, purslane, redmaids, ryegrass, sandbur, shepherdspurse, Texas panicum.

BENSULIDE, BETASAN, PREFAR   *Mfg: Stauffer*

Annual bluegrass, barnyard grass, crabgrass, deadnettle, fall panicum, foxtails, goosegrass, jungle rice, lambsquarters, pigweed, purslane, shepherdspurse, springletop, watergrass.

BENTAZON, BASAGRAN   *Mfg: BASF*

Arrowhead, beggarticks, bristly starbur, Canada thistle, cocklebur, day flower, duck salad, giant ragweed, hairy nightshade, jimsonweed, ladys thumb, lambsquarters, morningglory, pigweed, prickly sida, purslane, ragweed, redstem, river bulrush, shepherdspurse, smartweed, spike rush, spurred anoda, sunflower, velvetleaf, venice mallow, water plaintain, wild mustard, yellow nutgrass.

BIFENOX, MODOWN   *Mfg: Rhone Poulenc*

Annual bluegrass, annual sedge, barnyard grass, burnweed, carpetweed, chickweed, day flower, duck salad, hemp sesbania, jimsonweed, kochia, lambsquarters, morningglory, pigweed, purslane, redstem, smartweed, sprangletop, teaweed, velvetleaf, water hyssop.

BORATES   *Mfg: U.S. Borax*

Bermudagrass, bindweed, Canada thistle, dogbane, johnsongrass, Klamath weed, leafy spurge, most annual weeds, poison ivy, poison oak, toad flax, wild lettuce, white top, and annual weeds and grasses.

BROMACIL, HYVAR   *Mfg: DuPont*

Aster, bahiagrass, balsam apple, barnyard grass, bermudagrass, bluegrass, bouncing bet, bracken fern, bromegrass, bromesedge, broomsedge, bur buttercup, cheatgrass, China lettuce, Coloradograss, cottonweed, cottonwood, crabgrass, crowfoot, dallisgrass, dandelion, dogbane, dog fennel, elms, Flora's paintbrush, Florida pursley, foxtails, goldenrod, goosegrass, hackberry, henbit, Hialoa, horsetail, johnsongrass, jungle rice, lambsquarters, maples, mustard, nutgrass, oaks, orchardgrass, pangolagrass, paragrass, pigweed, pine, plaintain, poplar, prostate knotweed, puncturevine, purple top, purslane, quackgrass, ragweed, redbud, redtop, ryegrass, saltgrass, sandbur, sedge, sprangletop, sweetgum, sumac, torpedograss, turkey mullein, wild carrot, wild cherry, wild oats, willow wiregrass, willows, vasey grass.

BROMOXYNIL, BROMINIL, BUCTRIL

Mfg: Union Carbide and Rhone Poulenc

Annual morningglory, annual nightshade, bachelors button, Bassia, black mustard, burchervil, catch weed, corn cockle, cocklebur, corn chomomile, cow cockle, dog fennel, false flax, fanweed, fiddleneck, field pennycress, flixweed, fumitory, green smartweed, giant ragweed, gromwell, groundsel, henbit, Jacobs ladder, jimsonweed, knawel, kochia, lambsquarters, London rocket, mayors tail, miners lettuce, morningglory, peppergrass, pigweed, prickly lettuce, prostrate spurge, puncturevine, purple mustard, ragweed, redmaids, Russian thistle, saltbrush, marestail, shepherdspurse, silverleaf nightshade, smartweed, sunflower, tall waterhemp, tansy mustard, Tartary buckwheat, tarweed, tumble mustard, velvetleaf, wild buckwheat, wild mustard, wild radish, winter vetch, yellow rocket.

BUTYLATE, GENATE, SUTAN

Mfg: Stauffer

Barnyard grass, bermudagrass seedlings, crabgrass, fall panicum, foxtails, goosegrass, johnsongrass (seedling), purple nutgrass, sandbur, Texas panicum, volunteer sorghum, watergrass, wild cane, yellow nutgrass.

CACODYLIC ACID, PHYTAR, RADICATE

Mfg: Vineland and Vertac

Annual weeds.

CALAR

Mfg: Vineland

Bahiagrass, barnyard grass, carpet grass, centipedegrass, chickweed, dalisgrass, foxtails, goosegrass, johnsongrass, knotweed, lemongrass, lovegrass, nutgrass, sandbur, sedge, St. Augustine grass, witchgrass.

CANOPY

Mfg: DuPont

Cocklebur, common ragweed, copperleaf, Florida beggerweed, giant ragweed hemp sesbania, hophorn bean, jimsonweed, lambsquarters, morningglory, pigweed, prickly sida, purslane, sicklepod, smartweed, spotted spurge, sunflower, teaweed, velvetleaf.

CHLORAMBEN, AMIBEN

Mfg: Union Carbide

Barnyard grass, beggarweed, carpetweed, chickweed, coffee weed, crabgrass, fall panicum, foxtails, goosegrass, johnsongrass (from seed), kochia, lambsquarters, mustard, nightshades, pigweed, prickley sida, purslane, ragweed, Russian thistle, smartweed, spurge, stinkgrass, velvetleaf, venice mallow.

CHLORBROMURON, BROMEX, MALORAN

Mfg: CIBA-Geigy

Barnyard grass, bittercress, carelessweed, carpetweed, chickweed, cocklebur, conical catchfly, crabgrass, fall panicum, false flax, fiddleneck, Florida pursley, foxtail millet, foxtails, goosegrass, gromwell, groundsel, henbit, Jacobs ladder, jimsonweed, lambsquarters, Mayweed, mustard, nightshade, pennycress, pigweed, plaintains, prickley sida, purslane, ragweed, shepherdspurse, signal grass, smartweed, spring beauty, velvetleaf, witchgrass.

CHLORFENAC, FENAC

Mfg: Union Carbide

Alexander grass, alkali sida, barnyard grass, blueweed, bouncing bet, bur ragweed, chickweed, coontail, crabgrass, dandelion, dock, field bindweed, foxtail, goosegrass, henbit, Indian rush pea, johnsongrass (seedling), kochia, lambsquarters, leafy spurge, milfoil, morningglory, mouse-ear, pond weeds, poverty weed, nightshade, pigweed, plaintain, puncturevine, purslane, ragweed, Russian knapweed, Russian thistle, smartweed, Southern naiad, spike rush, spiny amaranth, sow thistle, Texas blueweed, watergrass, water stargrass.

CHLORIMURON-ETHYL, CLASSIC

Mfg: DuPont

Beggerweed, bristly starbur, cocklebur giant ragweed, hemp sesbania, jimsonweed, morningglory, pigweed, ragweed, sicklepod, smartweed, sunflower, wild poinsetta, yellow nutsedge.

CHLOROPROPHAM, CIPC, FURLOE

Mfg: PPG Industries

Annual bluegrass, annual ryegrass, barnyard grass, bitter cress, bromegrass, canary grass, carpetweed, chickweed, crabgrass, dodder, false flax, field sorrel, foxtail knotweed, henbit, ladys thumb, little barley, mustards, nightshade, purslane, rabbitsfoot grass, rattail fescue, ryegrass, shepherdspurse, Sibara spp., smartweed, stinkgrass, watergrass, wild oats, wild buckwheat, witchgrass.

CHLOROTHAL, DACTHAL

Mfg: SDS Biotech

Annual bluegrass, barnyard grass, browntop panicum, carpetweed, chickweed, crabgrass, creeping speedwell, dodder, Florida pursley, foxtail, goosegrass, groundcherry, johnsongrass (from seed), lambsquarters, lovegrass, pigweed, purslane, sandbur, spurge, Texas millet, witchgrass.

CHLOROXURON, TENORAN

Mfg: CIBA-Geigy

Annual bluegrass, barnyard grass, Bachiaria, carpetweed, chickweed, cocklebur, crabgrass, Florida pursley, goosegrass, groundsel, jimsonweed, lovegrass, lambsquarters, morningglory, nightshade, pigweed, puncturevine, ragweed, shepherdspurse, smartweed, spurry, velvetleaf, sicklepod, wild mustard.

CHLORSULFURON, GLEAN, TELAR   Mfg: DuPont

American strawberry, annual bluegrass, aster, Australian saltbrush, bedstraw, bouncing bet, buckhorn plaintain, burbeak-ckevil, bur clover,Canadian thistle, chickweed, cinquefoil, conial catchfly, curly dock, dandelion, dog fennel, false chamomile, false flax, fiddleneck, field pennycress, filaree, foxtails, goldenrod, gromwell, groundsel, henbit, hoarycress, horsetails, knotweed, kochia, ladys thumb, lambsquarters, London Rocket, mallow, milk thistle, miners lettuce, musk thistle, mustards, nettle, pepperweed, pigweed, pineapple weed, prickly lettuce, puncture vine, ragweed, red clover, Russian thistle, ryegrass, scouring rush, shepherdspurse, smartweed, sowthistle, speedwell, star thistle, sunflower, sweet clover, tansy mustard, tumble mustard, vetch, white clover, white cockle, wild buckwheat, wild carrot, wild parsnips, wild radish, wild turnip, yarrow.

COLLEGO   Mfg: Upjohn Co.

Northern joint vetch, curly indigo.

COPPER SULFATE AND   Mfg: Applied Biochemists,
ITS DERIVATIVES   Cities Services,
                          Kocide, Inc.,
                          Phelps-Dodge,
                          Sandoz and others

Algae, hydrilla and other aquatic weeds.

CYCLOATE, RO-NEET   Mfg: Stauffer

Alligator weed, annual bluegrass, annual rye, black nightshade, crabgrass, foxtails, goosefoot, hairy nightshade, henbit, lambsquarters, nettle, purple nutgrass, purslane, red root pigweed, shepherdspurse, stinging nettle, volunteer barley, watergrass, wild oats, yellow nutgrass.

CYANAZINE, BLADEX   Mfg: DuPont

Annual bluegrass, annual morningglory, annual ryegrass, annual sedge, barn-yard grass, buffalobur, bullgrass, buttercup, carpetweed, cheatgrass, chickweed, cocklebur, corn spurry, crabgrass, curly dock, downybrome, fall panicum, false flax, fescue, Flora's paintbrush, Florida pursley, foxtails, galinsoga, goosegrass, groundcherry, groundsel, henbit, Indian lovegrass, jimsonweed, johnsongrass, jungle rice, knotweed, kochia, ladys thumb, lambsquarters, mallow, Mayweed, mustard, pigweed, pineapple weed, plaintain, poorjoe, prickly lettuce, purslane, ragweed, Russian thistle, shepherdspurse, smartweed, spinysida, spurge, stinkgrass, tarweed, teaweed, velvetleaf, volunteer wheat, wild buckwheat, wild turnip, wild radish, wild oats, witchgrass.

DALAPON, BASAPON   *Mfg: Dow and BASF*

Alexander grass, amargo grass, all annual grasses, bermudagrass, bluegrass, burweed, carpet grass, cattails, crabgrass, cutgrass, dallisgrass, foxtail, Gamalote grass, giant foxtail, guinea grass, johnsongrass, Kikuya grass, maiden cane, pangola grass, paragrass phragmites, poverty grass, quackgrass, rushers, sedges, watergrass.

DESMEDIPHAM, BETANEX   *Mfg: Nor-Am Ag. Products*

Chickweed, coast fiddleneck, goosefoot, groundcherry, kochia, lambsquarters, London rocket, mustard, nightshade, pigweed, ragweed, shepherdspurse, sow thistle, wild buckwheat.

DEVINE   *Mfg: Abbott Labs*

Strangler or milkweed vine.

DI ALLATE, AVADEX   *Mfg: Monsanto*

Wild oats.

DICAMBA, BANVEL   *Mfg: Velsicol*

Alfalfa, alligator weed, annual morningglory, arrowhead, ash, aspen, asters, balloonvine, basswood, bedstraw, bitter dock, blackberry, bladder champion, bloodweed, blueball, blueweed, bracken fern, buckrush, buffalo bur, bullnettle, bur clover, Canada thistle, Carolina geranium, carpetweed, cattail, cedar, cheatgrass, cherry, chickweed, chickory, chinquapin, clovers, cocklebur, corn chamomile, corn cockle, cottonwood, cow cockle, crabgrass, creosote bush, croton, cucumber tree, dalmation toad flax, dandelion, dewberry, dock, dog fennel, dogwood, dragonhead, Eastern persimmon, elm, English daisy, fiddleneck, field bindweed, field peppergrass, field pennycress, foxtails, french weed, frongbit, giant ragweed, goatsbeard, goldenrod, grape, gromwells, hardwood trees, hawkweed, hemp dogbane, hempnettle, henbit, hickory, honeysuckle, hornbean, horsemint, horseweed, jimsonweed, knawel, knotweed, kochia, kudza, ladys thumb, lambsquarters, lawn burweed, leafy spurge, locust, London Rocket, lote, lupines, mallow, maples, marsh thistle, Mayweed, milkweeds, morningglory, multiflora rose, musk thistles, mustards, nettles, nightflowering catchfly, nightshade, nutgrass, oak, parrotfeather, pepperweed, persimmon, pickerelweed, pigweed, pine, pingue, poison ivy, poison oak, pokeeweed, poorjoe, popular, poverty weed, puncturevine, purslane, rabbit brush, ragweed, rattail fescue, rattlebush, red vine, ripgut brome, rough stumpweed, Russian knapweed, Russian thistle, sagebrush, sassafras, service berry, sesbania, sheep sorrel, shepardspurse, silverleaf nightshade, slender spike rush, small leafsida, smartweed, snakeweed, sourwood, sow thistle, Spanish nettle, spicebush, spikeweed, spurge spurry, star thistle, stitchwort, stinkweed, sumac, sunflower, swamp smartweed, Sycamore, Tansy ragwort, tarbrush, tarweed, teasel, thorn apple, thorn berry, tievine, trumpet creeper, velvetleaf, vetch, watergrass, water hemlock, water hemp, water hyacinth, water pennywort, water primrose, white cockle, wild buckwheat, wild carrot, wild garlic, wild onion, wild radish, willow, witch hazel, wood sorrel, worm wood, yarrow, yaupon, yellow thistle, yucca.

DICHLORBENIL, CASORON, NOROSAC

Mfg: PBI Gordon

Annual bluegrass, artemisia, bindweed, bluegrass, bull thistle, camphorweed, Canada thistle, carpetweed, chora, chickweed, citron melon, coffee weed, coontail, crabgrass, cudweed, dandelion, dock, dog fennel, elodea, evening primrose, false dandelion, fescue, fiddleneck, Florida purslane, foxtail, gisekia, goosefoot, groundsel, henbit, horsetail, knotweed, lambsquarters, leafy spurge, Maypops, milkweed vine, miners lettuce, naiads, natalgrass, old witchgrass, orchardgrass, peppergrass, pigweed, pineapple weed, plaintain, pond weeds, purslane, quackgrass, ragweed, red deadnettle, redroot, rosary pea, Russian knapweed, Russian thistle, sheep sorrel, shepherdspurse, smartweed, Spanish needles, spurge, teaweed, Texas panicum (Harrahgrass), timothy, water milfoil, wild artichoke, wild aster, wild barley, wild carrot, wild mustard, wild radish, yellow rocket, yellow wood sorrel.

DICHLORPROP, 2,4-DP

Mfg: Union Carbide

Most annual broadleaf weeds.

DICLOFOP-METHYL, HOELON

Mfg: American Hoechst

Annual ryegrass, barnyard grass, broadleaf signal grass, canarygrass, crabgrass, downy brome, fall panicum, foxtails, Persian darnel, ripgut brome, volunteer corn, wild oats, witchgrass.

DIETHATYL-ETHYL, ANTOR

Mfg: Nor-Am

Annual ryegrass, annual sow thistle, barnyard grass, black nightshade, broadleaf signal grass, canary grass, crabgrass, fall panicum, foxtail millet, foxtails, groundcherry, johnsongrass (seedlings), jungle rice, pigweed, shepherdspurse, sprangletop, wild oats, witchgrass.

DIFENZOQUAT, AVENGE   *Mfg: American Cyanamid*

Wild oats.

DIMETHAZONE, COMMAND   *Mfg: FMC Corp.*

Barnyardgrass, broadleaf signalgrass, cocklebur, Florida beggerweed, Florida pusley, foxtails, goosegrass, jimsonweed, johnsongrass, lambsquarters, panicums, pitted morningglory, prickly sida, purslane, ragweed, sandbur, smartweed, southwestern cupgrass, spurge, tropic croton, velvetleaf, venice mallow, wooly cupgrass.

DIPHENAMID, ENIDE   *Mfg: Nor-Am*

Annual bluegrass, annual sedge, barnyard grass, carpetweed, cheatgrass, chickweed, corn spurry, crabgrass, crowfoot grass, evening primrose, fall panicum, Florida pursley, foxtail, groundsel, goosegrass, johnsongrass, knotgrass, knotweed, lambsquarters, peppergrass, pigweed, purslane, red sorrel, ryegrass, sandbur, shepherdspurse, smartweed, spiny amaranth, stinkgrass, Thymeleaved sandwort, white clover, wild oats, witchgrass.

DIPROPETRYN, SANCAP   *Mfg: CIBA-Geigy*

Barnyard grass, crabgrass, groundcherry, lambsquarters, pigweeds, Russian thistle, sandbur, spurred anoda.

DIQUAT   *Mfg: Chevron*

Algae, bladderwort, cattails, coontail, Elodea, most small annual weeds, naiad, pennywort, pond weeds, salvinia, water hyacinth, water lettuce, water milfoil.

DIURON, KARMEX   *Mfg: DuPont and others*

Ageratum, Amsineka (fiddleneck), annual groundcherry, annual lovegrass, annual morningglory, annual ryegrass, annual smartweed, annual sow thistle, annual sweet vernalgrass, barnyard grass, bluegrass, buttonweed, chickweed, crabgrass, corn speedwell, corn spurry, day flower, dog fennel, foxtails, gromwell, groundsel, hawkbeard, horseweed, johnsongrass (from seed), knawel, kochia, Kyllinga, lambsquarters, marigold, Mexican clover, mustard, orchard grass, pennycress, peppergrass, pigweed, pineapple weed, pokeweed, purslane, rabbit tobacco, ragweed, rattail fescue, red sprangletop, ricegrass, sandbur, shepherdspurse, Spanish needles, tansy mustard, velvetgrass, watergrass, wild buckwheat, wild lettuce, wild mustard, wild radish.

DSMA

Mfg: Fermenta and Vineland

Bahiagrass, barnyard grass, chickweed, cocklebur, dallisgrass, goosegrass, johnsongrass, nutgrass, puncturevine, ragweed, sandbur, wood sorrel.

DYANAP, ANCRACK

Mfg: Uniroyal and others

Barnyard grass, beggarweed, bindweed, black nightshade, carpetweed, chickweed, cocklebur, crabgrass, cupgrass, Florida pursley, foxtails, galinsoga, goosegrass, groundcherrry, jimsonweed, lambsquarters, morningglory, mustards, pigweed, purslane, ragweed, sandbur, seedling johnsongrass, shepherdspurse, smartweed, sprangletop, stinkgrass, sunflower, teaweed, velvetleaf, watergrass, windmill grass.

ENDOTHAL, AQUATHOL, HYDROTHOL

Mfg: Pennwalt

Algae, annual bluegrass, barnyard grass, bassweed, bromes, black medic, bullgrass, bur clover, bur reed, carrot weed, cheatgrass, chickweed, clovers, coontail, cranesbill, dichrondra, foxtail, filaree, goathead, henbit, hydrilla, knotweed, kochia, little barley, lespedeza, naiad, oxalis, pigweed, pond weeds, purslane, ragweed, rescuegrass, ryegrass, shepherdspurse, smartweed, Texas blueweed, vetch, volunteer barley, veronica, watergrass, water milfoil, water stargrass, widgeon grass, wild buckwheat.

EPTC, EPTAM, GENEP

Mfg: Stauffer and PPG

Annual bluegrass, annual morningglory, annual ryegrass, barnyard grass, bermudagrass, black nightshade, carpetweed, chickweed, corn spurry, crabgrass, fiddleneck, Florida purslane, foxtail, goosegrass, hairy nightshade, henbit, johnsongrass, lambsquarters, lovegrass, fall panicum, mugwort, nettleleaved goosefoot, prostrage pigweed, purple nutgrass, purslane, quackgrass, ragweed, red root pigweed, rescuegrass, sandbur, shattercane, shepherdspurse, signal grass, tumbling pigweed, volunteer grains, watergrass, wild oats, yellow nutgrass.

ERADICANE

Mfg: Stauffer

Annual bluegrass, annual morningglory, annual ryegrass, barnyard grass, bermudagrass, carpetweed, chickweed, corn spurry, crabgrass, deadnettle, fall panicum, fiddleneck, Florida purslane, foxtails, goosefoot, goosegrass, johnsongrass, lambsquarters, lovegrass, nightshade, nutgrass, pigweed, puncturevine, purslane, quackgrass, ragweed, rescuegrass, sandbur, signal grass, shepherdspurse, teaweed, Texas panicum sicklepod, velvetleaf, volunteer grains, wild cane, wild oats, wooly cupgrass.

ETHALFURALIN, SONALAN   *Mfg: Elanco*

Annual bluegrass, barnyard grass, carpet grass, catchfly, chickweed, crabgrass, fall panicum, fiddleneck, Florida pursley, foxtails, foxtail millet, goosegrass, groundcherry, henbit, johnsongrass, jungle rice, kochia, lambsquarters, nightshade, pigweed, purslane, ragweed, rock purslane, Russian thistle, ryegrass, shattercane, signal grass, Texas panicum, wild buckwheat, wild mustard, wild oats, witchgrass.

ETHOFUMESATE, NORTRON,   *Mfg: Nor-Am Chemicals*
PROGRASS

Annual bluegrass, barnyard grass, canary grass, chickweed, crabgrass, downey brome, fescue, foxtails, kochia, ladys thumb, lambsquarters, mannagrass, nightshade, pigweed, puncturevine, purslane, Russian thistle, smartweed, softchess, velvetgrass, volunteer barley, volunteer wheat, wild buckwheat, wild oats.

FINESSE   *Mfg: DuPont*

Annual bluegrass, annual ryegrass, bedstraw, broadleaf dock, buckwheat, bur buttercup, Canadian thistle, chuckweed, conical catchfly, corn groundsel, corn spurry, cow cockle, dovefoot geranium, false chamomile, fiddleneck, filaree, field pennycress, flexweed, foxtails, groundsel, hemp nettle, henbit, Jacobs ladder, knotweed, kochia, ladysthumb, lambsquarters, little bittervress, mayweed, pigweed, pineapple weed, prickly lettuce, prickly poppy, purslane, Russian thistle, shepherdspurse, smartweed, sowthistle, speedwell, tansy mustard, vetch, white cockle, wild buckwheat, wild carrot, wild radish.

FLOUMETURON, COTORAN   *Mfg: CIBA-Geigy*

Barnyard grass, Brachiaria, buttonweed, cocklebur, crabgrass, crowfoot grass, fall panicum, Florida pursley, foxtail, goosegrass, groundcherry, jimsonweed, lambsquarters, morningglory, pigweed, prickley sida, puncturevine, purslane, ragweed, ryegrass, sesbania, sicklepod, smartweed, tumbleweed.

FLUAZIPROP-BUTYL, FUSILADE   *Mfg: ICI Americas*

Barnyard grass, bermudagrass, crabgrass, cupgrass, fall panicum, foxtails, goosegrass, itchgrass, johnsongrass, jungle rice, quackgrass, red rice, sandbur, shattercane, signal cane, Texas panicum, volunteer corn, wild oats, wild proso millet, wirestem muley.

FLURIDONE, SONAR   *Mfg: Elanco*

Bladderwort, common coontail, common duckweed, creeping waterprimrose, egeria, elodea, hydrilla, naiad, paragrass, pondweed, reed canarygrass, spatterdock, water lily, water milford, water purslane.

FOSAMINE, KRENITE   *Mfg: DuPont*

Alder, American alder, ash, black gum, blackberry, cottonwood, Eastern white pine, elm, hawthorn, hickory, locust, maple, multiflora rose, oaks, pines, poplar, sassafras, sumac, sweetgum, sycamore, thimbleberry, tree of heaven, vine maple, wild cherry, wild grape, wild plum, willow.

GEMNI   *Mfg: DuPont*

Carpetweed, cocklebur, common ragweed, Florida beggarweed, giant ragweed, hemp, jimsonweed, lambsquarters, morningglory, mustards, pigweed, purslane, sesbania, sicklepod, smartweed, sunflower, teaweed, velvetleaf.

GLYPHOSATE, RODEO, ROUNDUP   *Mfg: Monsanto*

Alfalfa, annual bluegrass, bahiagrass, barley, barnyardgrass, bassia, bermudagrass, berries, bindweed, bluegrass, broken fern, brome, camel thorne, Canada thistle, canary grass, cattails, chickweed, clovers, cocklebur, common mullein, crabgrass, curly dock, dallisgrass, dandelion, downybrome, falseflax, fescues, fiddleneck, field bindweed, flaxleaf fleabane, foxtails, giant cutgrass, giant ragweed, guinea grass, groundsel, hemp dogbane, honeysuckle, horsenettle, horseradish, horseweed, Jerusalem artichoke, johnsongrass, Kentucky bluegrass, kikuyugrass, knapweed, kochia, kudzu, lambsquarters, lantana, London rocket, maiden cane, maples, milkweed, multiflora rose, orchardgrass, Napiergrass, nutsedge, oaks, panicum, paragrass, pennycress, phragmites, pigweed, poison ivy, poison oak, prickly lettuce, quackgrass, ragweed, red canary grass, Russian thistle, ryegrass, sandbur, shattercane, shepherdspurse, smartweed, sowthistle, Spanish needles, shatter dock, sunflower, Texas blueweed, timothy, torpedograss, trumpet creeper, tules, vasey grass, velvetleaf, volunteer corn, volunteer wheat, wheatgrass, wild oats, wild oats, wild sweet potatoes, willows, wirestem muley.

HEXAZINONE, VELPAR   *Mfg: DuPont*

Annual bluegrass, annual ryegrass, bahiagrass, barnyard grass, bermudagrass, bindweed, bluegrass, bouncing bet, broomsedge, bromegrass, burdock, camphorweed, Canada thistle, cheatgrass, chickweed, clovers, cocklebur, crabgrass, crown vetch, dallisgrass, dandelion, henbit, dewberry, dock, dog fennel, dogbane, fescue, fiddleneck, field pennycress, filaree, fingergrass, fleabane, foxtail, foxtail barley, fuffelgrass, goatsbeard vine, goldenrod, groundsel, guinea grass, heath aster, honeysuckle, ivyleaf speedwell, kochia, lambsquarters, lantana, lespedeza, London rocket, marestail, Mexican tea, milkweed, miners lettuce, mustards, notchgrass, nutsedge, orchardgrass, oxalia, paragrass, pigweed, plaintain, prickly lettuce, purslane, quackgrass, ragweed, ryegrass, seedling alfalfa, shepherdspurse, smartweed, smutgrass, Spanish needle, spurge, star thistle, sweet clover, tansy mustard, trumpet creeper, vasey grass, wild blackberry, wild carrot, wild oats, wild parsnip, yellow rocket.

IMAZAPYR, ARSENAL   Mfg: American Cayanamid

Ash, bahiagrass, beardgrass, bermudagrass, big bluestem, bindweed, blackberry, bluegrass, bull thistle, burdock, camphor weed, Canada thistle, Carolina geranium, carpet weed, cattail, cheat, cherry, chickweed, churchgrass, clover, cocklebur, crabgrass, curly dock, dallisgrass, dandelion, dewberry, dog fennel, dog weed, downy brome, filaree, fleabane, foxtails, goldenrod, goosefoot, goosegrass, green briar, guineagrass, hawthorne, hickory, hoary vervain, honey locust, honey suckle, johnsongrass, kochia, kudzu, lambsquarters, lespedeza, lovegrass, mallow, milkweed, maple, marestail, miners lettuce, morningglory, mulberry, mullein, multiflore roses, mustard, oak, opossum grape, ox-eye daisy, panicum, paragrass, pepperweed, pigweed, plaintain, poison ivy, pokeweed, popular, prairie threeawn, primrose, privet, purslane, quackgrass, ragweed, redvine, ryegrass, sand dropseed, sandspur, sassafras, signalgrass, silverleaf, nightshade, smartweed, smooth brome, sorrel, sow thistle, sumac, sweet clover, tall fescue, Texas thistle, timothy, torpedograss, trumpet creeper, vaseygrass, Virginia creeper, wild barley, wild buckwheat, wild carrot, wild lettuce, wild mustard, wild oats, wild parsnip, wild turnip, willow, wirestem muhly, witchgrass, willyleaf bursage, yellow star thistle, yellow wood-sorrel.

IMAZAQUIN, SCEPTER   Mfg: American Cyanamid

Cocklebur, Florida pusley, foxtails, giant ragweed, jimsonweed, lambsquarters, morningglory, mustard, nightshade, pigweed, ragweed, seedling johnsongrass, smartweed, teaweed, velvetleaf, Venice mallow, wild poinsetta, wild sunflower.

ISOPROPALIN, PAARLAN   Mfg: Elanco

Barnyard grass, carelessweed, crabgrass, foxtails, goosegrass, johnsongrass (from seed), lambsquarters, pigweed, purslane, ryegrass.

KROVAR   Mfg: DuPont

Bahiagrass, barnyard grass, bermudagrass, bouncing bet, chickweed, crabgrass, dogbane, filaree, fleabane, Florida pursley, foxtails, groundsel, horseweed, johnsongrass, jungle rice, lambsquarters, natalgrass, nightshade, nutsedge, pangolagrass, pigweed, paragrass, pineapple weed, puncturevine, purslane, ragweed, saltgrass, sandbur, seedling sow thistle, Spanish needles, torpedograss, wild lettuce, wild mustard.

LINURON, LOROX   *Mfg: DuPont*

Amsineka, annual ryegrass, buttonweed, canary grass, carpetweed, chickweed, crabgrass, dog fennel, fall panicum, Florida pursley, foxtails, galinsoga, goosefoot, goosegrass, gromwell, groundsel, knawel, lambsquarters, morningglory, mustard, nettleleaf, pigweed, prickley sida, purslane, ragweed, rattail fescue, sesbania, sicklepod, smartweed, Texas panicum, velvetleaf, watergrass, wild buckwheat, wild radish.

MCPA   *Mfg: Numerous*

Arrowhead, beggarsticks, bullrush, burhead, burdock, buttercups, Canada thistle, cocklebur, curly indigo, dandelion, dragonhead mint, goatsbeard, hemp-nettle, kochia, lambsquarters, marsh elder, meadow buttercup, mustards, night-shade, peppergrass, pigweed, pennycress, plaintain, poison hemlock, punc-turevine, purslane, ragweed, redstem, Russian thistle, sedge, shepherdspurse, sow thistle, stinkweed, stinging nettle, sunflowers, vetch, water plaintain, white top, wild radish, winter cress, water hyssop, yellow rocket.

MCPB   *Mfg: Union Carbide and*
                                             Rhone Poulenc

Canada thistle, fanweed, lambsquarters, morningglory, nightshade, pigweed, smartweed, sow thistle.

MECOPROP, MCPP   *Mfg: Numerous*

Chickweed, cleavers, clover, dichronda, field pennycress, ground ivy, knot-weed, lambsquarters, mustard, pigweed, plaintain, ragweed, shepherdspurse, wild radish.

MEFLUIDIDE, EMBARK   *Mfg: 3M Co.*

Johnsongrass, shattercane, volunteer corn, volunteer wheat.

METOLACHLOR, DUAL, PENNANT   *Mfg: CIBA-Geigy*

Barnyard grass, carpetweed, crabgrass, crowfoot grass, cupgrass, fall panicum, Florida pursley, foxtail millet, foxtails, galinsoga, goosegrass, nightshade, pigweed, red rice, shattercane, signal grass, witchgrass, yellow nutsedge.

METRIBUZIN, LEXONE, SENCOR   *Mfg: Mobay & DuPont*

Ageratum, Alexandergrass, alkali mallow, annual bluegrass, bedstraw, beg-garweed, bindweed, buffalobur, bur clover, canary grass, Carolina geranium, carpetweed, cheatgrass, chickweed, cocklebur, coffeeweed, conical catchfly, corn cockle, corn speedwell, crabgrass, crowfoot grass, curly dock, day flower, dead nettle, dog fennel, euphorbia, fiddleneck, field pennycress, filaree, fireweed, floras paintbrush, Florida purslane, foxtails, goosefoot, goosegrass, gromwell, guineagrass, Haoloe koa, henbit, Hialsa, Hilgihila, hop clover, Jacobs ladder, jimsonweed, knotweed, lambsquarters, London rocket, mad wort, malva, marestail, Mexican weed, miners lettuce, mustards, panicum, parsleypiert, pigweed, plushgrass, prickley sida, purslane, rabbits foot grass, ragweed, rattlepot, rice grass, Richardia, seedling johnsongrass, sesbania, shepherdspurse, sicklepod, signal grass, six weeks gramagrass, smallflowered buttercup, smartweed, sow thistle, speedwell, spiny amaranths, spur weed, spurge, tarweed, toadflax, velvetleaf, venice mallow, white clover, winter oats, wiregrass.

METSULFURON-METHYL, ALLY   Mfg: DuPont

Bittercress, bur buttercup, Canadian thistle, chickweed, conical catchfly, corn cockle, corn gromwell, cow cockle, dog fennel, false chamomile, fiddleneck, field pennycress, filaree, flixweed, groundsel, knotweed, kochia, lambsquarters, mayweed, miners lettuce, mustards, pigweed, prickly lettuce, purslane, Russian thistle, shepherdspurse, smallseed toadflax, smartweed, sowthistle, sunflower, tansy mustard, treacle mustard, tumble mustard, waterpod, wild buckwheat.

MOLINATE, ORDRAM   Mfg: Stauffer

Annual sedges, barnyard grass, Brachiaria, crabgrass, day flower, red rice, spike rush, watergrass.

MSMA   Mfg: Fermenta, Vertac and Vineland

Bahiagrass, barnyard grass, chickweed, cocklebur, crabgrass, dallisgrass, johnsongrass, foxtails, mustard, nutgrass, pigweed, puncturevine, ragweed, sandbur, tules, wild oats, wood-sorrel.

NAPROPAMIDE, DEVRINOL   Mfg: Stauffer

Annual bluegrass, barnyard grass, bromes, carpetweed, cheatgrass, crabgrass, cupgrass, chickweed, fall panicum, fiddleneck, filaree, foxtails, goosegrass, groundsel, knotweed, lambsquarters, little mallow, mallow, panicum, pigweed, pineapple weed, prickley lettuce, purslane, ragweed, ripgut brome, ryegrass, sandbur, seedling johnsongrass, soft chess, sow thistle, sprangletop, stinkgrass, wild barley, wild oats, witchgrass.

NAPTALAN, ALANAP   Mfg: Uniroyal Inc.

Barnyard grass, bindweed, carpetweed, chickweed, cocklebur, crabgrass, cupgrass, foxtails, galinsoga, goosegrass, groundcherry, johnsongrass (from seed), lambsquarters, mustard, pigweed, purslane, ragweed, sandbur, shepherdspurse, sprangletop, stink grass, velvetleaf, watergrass, windmill grass.

NORFLURAZON, EVITAL,   Mfg: Sandoz
SOLICAM, ZORIAL

Annual bluegrass, annual bursage, annual sedge, bahiagrass, barnyardgrass, bog rush, broomsedge, Carolina geranium, carpetweed, cheat, cheeseweed, chickweed, crabgrass, cudweed, cupgrass, dog fennel, downy brome, fall panicum, false dandelion, feather fingergrass, fescues, fiddleneck, filaree, Florida pursley, foxtails, goosegrass, johnsongrass seedlings, London rocket, needlegrass, nutgrass, panicum, pigweed, pineapple weed, prickly sida, poverty grass, puncturevine, purslane, redmaids, redroot, redtopgrass, rice cutgrass, ripgut brome, Russian thistle, ryegrass, salt grass, shepherdspurse, sicklegrass, signalgrass, six weeks gramagrass, smartweed, smokegrass, spikerush, spurge, spurred anoda, sowthistle, stargrass, summergrass, switchgrass, tall fescue, Tropic croton, velvetleaf, Virginia pepperweed, wild barley, wild buckwheat, wild geranium, wild onion, witchgrass, woolgrass.

ORYZALIN, SURFLAN   Mfg: Elanco

Annual bluegrass, annual ryegrass, barnyard grass, bitter cress, Brachiaria, browntop panicum, carpetweed, chickweed, crabgrass, creeping wood-sorrel, crowfoot grass, cupgrass, fall panicum, fiddleneck, filaree, Florida pursley, foxtails, goosegrass, groundsel, henbit, horseweed, johnsongrass (from seed), jungle rice, knotweed, lambsquarters, little barley, London rocket, lovegrass, oxalis, panicums, pigweed, prostrate spurge, puncturevine, purslane, redmaids, red rice, sandbur, shepherdspurse, sow thistle, sprangletop, spurge, wild oats, witchgrass.

OXADIAZON, RONSTAR   Mfg: Rhone Poulenc

Annual bluegrass, barnyard grass, bittercress, carpetweed, crabgrass, common day flower, crabgrass, duck salad, evening primrose, fall panicum, Florida pursley, galinsoga, golden ragwort, goosegrass, groundsel, jungle rice, lambsquarters, liverwort, niruri, oxalis, Panama paspalum, petty spurge, pigweed, poa annua, prostrate spurge, purslane, red sprangletop, redstem, sensitive plant, smartweed, sowthistle, speedwell, spiny amaranths, spotted catsears, spurge, stinging nettle, swine cress, Texas panicum, water hyssop, white stemmed filaree, wood-sorrel, yellow wood-sorrel.

OXYFLUORFEN, GOAL   *Mfg: Rohm and Haas*

Annual bluegrass, annual morningglory, balsum apple, barnyardgrass, bed-straw, bittercress, burclover, camphorweed, canarygrass, carpetweed, cheeseweed, chickweed, clovers, cockelbur, corn spurry, crabgrass, cudweed, cupgrass, evening primrose, fall panicum, fiddleneck, filaree, fireweed, flexweed, foxtails, goosegrass, groundcherry, groundsel, hemp sesbania, hen-bit, horseweed, jimsonweed, knotweed, ladysthumb, lambsquarters, lanceleaf sage, London rocket, malva, mayweed miners lettuce, mustards, nettle, night-shade, oxalis, pepperweed, pigweed, pineapple, prickly lettuce, puncture vine, purslane, red maids, red orach, red sandspurry, red sorrel, ripgut brome, Russian thistle, scarlet pimpernel, shepherdspurse, smartweed, sow thistle, speedwell, spurge, sweet clover, tansy mustard, teaweed, tropic croton, velvetleaf, wild buckwheat, wild oats, wild poinsetta, wild radish, witchgrass, witchweed, yellow sorrel.

PARAQUAT, GRAMOXONE   *Mfg: ICI*

Annual grasses, annual ryegrass, bermudagrass, bluegrass, bur clover, ch-eatgrass, chickweed, crabgrass, filaree, groundsel, johnsongrass, morningglory, mettle, pigweed, plaintain, puncturevine, purslane, red clover, shepherdspurse, thistles, volunteer barley, wild mustard, wild mustard, wild oats, wild radish.

PEBULATE, TILLAM   *Mfg: Stauffer*

Barnyard grass, bermudagrass, blackeyed Susan, crabgrass, deadnettle, Florida pursley, foxtails, goosefoot, goosegrass, hairy nightshade, lambsquarters, pigweed, purple nutgrass, purslane, shepherdspurse, signal grass, wild oats, yellow nutgrass.

PENDIMETHALIN, PROWL   *Mfg: American Cyanamid*

Annual bluegrass, annual spurge, barnyard grass, carpetweed, chickweed, crabgrass, crowfoot grass, fiddleneck, filaree, Florida pursley, foxtails, goosegrass, henbit, itchgrass, johnsongrass seedlings, jungle rice, knotweed, kochia, lambsquarters, London rocket, lovegrass, panicums, pigweed, prosom-illet, puncturevine, purslane, shepherdspurse, signal grass, smartweed, spran-gletop, spurge, velvetleaf, witchgrass, wolly cupgrass.

PHENMEDIPHAM, BETANAL   *Mfg: Nor-Am Ag. Products*

Chickweed, fiddleneck, goosefoot, green foxtail, groundcherry, kochia, lambs-quarters, London rocket, mustard, nightshade, pigeon grass (yellow foxtail), pigweed, purslane, ragweed, shepherdspurse, sow thistle, wild buckwheat.

PICLORAM, TORDON, GRAZON   *Mfg: Dow*

Ailanthus, alder, artichoke thistle, ash, aspen, balsam, bindweed, birch, bitter-weed, black gum, blackberry, bouncing bet, bracken fern, brambles, broomweed, bursage, button brush, bur ragweed, Canadian thistles, cedar, chicory, chokeberry, cholla cacti, conifers, cypress, dandelion, dock, dogwood, elderberry, elm, field bindweed, fir, fleabane, goldenrod, golfberry, grose, groundcherry, gums, hawthorn, hemlock, hickory, honeysuckle, horse nettle, junipers, knapweed, kudzu, larkspur, leafy spurge, loco weeds, locusts, lupines, manzanita, maple, mesquite, milkweed, most broadleaf weeds, mulberry, musk thistle, oak, persimmon, pigweed, pine, plaintain, poison oak, poverty weed, prickley lettuce, prickley pear, rabbit brush, ragweed, redband, Russian knapweed, sassafras, scotch thistle, skeleton weed, snakeweed, sourwood, sow thistle, spruce, star thistle, sumac, sunflower, sweet clover, tansy ragwort, tansy thistle, toad flax, tulip poplar, vetch, wild carrot, wild cherry, wild grapes, wild rose, willows.

PROMETONE, PRAMITOL   *Mfg: CIBA-Geigy*

Annual weeds, bermudagrass, bindweed, downy bromegrass, goldenrod, goosegrass, johnsongrass, oatgrass, plaintain, puncturevine, quackgrass, wild carrot.

PROMETRYNE, CAPAROL   *Mfg: CIBA-Geigy*

Annual morningglory, barnyard grass, carpetweed, carelessweed, cranes bill, cheatgrass, cocklebur, coffee weed, crabgrass, chickweed, fiddleneck, field pennycress, fleabane, Florida pursley, foxtails, galinsoga, goosegrass, ground-cherry, groundsel, horseweed, jimsonweed, jungle rice, knotweed, kochia, ladys thumb, lambsquarters, little barley, malva, mayweed, mustards, nightshade, pigweed, pineapple weed, prickley sida, purslane, ragweed, rattail fescue, Russian thistle, sandbur, shepherdspurse, signal grass, skeleton weed, smartweed, spurred anoda, sunflower, teaweed, velvetleaf, velvetgrass, water-grass, water hemp.

PRONAMIDE, KERB   *Mfg: Rohm & Haas*

Annual morningglory, barnyard grass, bentgrass, bluegrass, bromes, burning nettle, canary grass, carpetweed, cheat, chickweed, crabgrass, dodder, fall panicum, fescues, foxtail barley, foxtails, goosefoot, goosegrass, henbit, knot-weed, lambsquarters, London rocket, lovegrass, mustards, nettles, nightshade, orchardgrass, purslane, quackgrass, ryegrass, sheep sorrel, shepherdspurse, smartweed, velvetgrass, volunteer grain, volunteer tomatoes, wild oats, wild radish.

PROPACHLOR, RAMROD   *Mfg: Monsanto*

Annual ryegrass, barnyard grass, carpetweed, crabgrass, fall panicum, Florida pursley, giant foxtail, goosegrass, green foxtail, groundsel, lambsquarters, pigweed, purslane, ragweed, sandbur, shepherdspurse, smartweed, watergrass, wild buckwheat, yellow foxtail.

PROPANIL, STAM, STAMPEDE   *Mfg: Rohm & Haas*

Alligator weed, barnyard grass, Brachiaria, burhead, crabgrass, curly dock, curly indigo, foxtails, goosegrass, gulf cockspur, Hoorahgrass, jointed sedge, kochia, lambsquarters, Mexican weed, paragrass, pigeongrass, pigweed, redstem, redweed, spearhead, spike rush, tall indigo, Texas millet, watergrass, wild buckwheat, wild mustard, wiregrass, wooly croton.

PROPAZINE, MILOGARD   *Mfg: CIBA-Geigy*

Annual morningglory, carpetweed, foxtail, lambsquarters, most annual broadleaves, most annual grasses, pigweed, ragweed, red rice, smartweed, velvetleaf.

PROPHAM, CHEM-HOE, IPC   *Mfg: PPG Industries*

Annual bluegrass, annual ryegrass, bromegrass, bulbous bluegrass, canary grass, cheatgrass, chickweed, downybrome, downeychess, foxtail, jointed goatgrass, pigeon grass, rabbitsfoot grass, rattail fescue, small nettle, velvetgrass, volunteer grains, wild oats.

PRYRAZON, PYRAMIN   *Mfg: BASF*

Buffalobur, chickweed, fanweed, fiddleneck, goosefoot, groundcherry, groundsel, henbit, knotweed, lambsquarters, London rocket, mayweed, mustard, nettleleaf, nightshade, pennycress, pigweed, prickley lettuce purslane, ragweed, shepherdspurse, smartweed, sow thistle, wild buckwheat, wild carrot.

ROUT   *Mfg: Sierra Chemical Co.*

Annual bluegrass, barnyardgrass, bittercress, chickweed, crabgrass, dandelion, fall panicum, foxtails, goosegrass, lambsquarters, oxalis, pigweed, purslane, shepherdspurse, sow thistle, spurge.

SETHOXYDIM, POAST   *Mfg: BASF*

Barnyard grass, bermudagrass, crabgrass, cupgrass, foxtails, goosegrass, itchgrass, johnsongrass, jungle rice, lovegrass, panicums, orchardgrass, quackgrass, red rice, ryegrass, signal grass, sprangletop, tall fescue, volunteer cereals, volunteer corn, wild cane, wild oats, wild proso millet, wirestem muhly, witchgrass, wooly cupgrass.

SIDURON, TUPERSAN   *Mfg: DuPont*

Crabgrass, foxtail, watergrass.

SIMAZINE, AQUAZINE, PRINCEP   *Mfg: CIBA-Geigy and others*

Algae (unicellular and filamentous), alyssum, amaranths, annual bluegrass, annual morningglory, annual ryegrass, barnyard grass, Brachiaria, bur clover, burdock, Canada thistle, carelessweed, carpetweed, cheatweed, chickweed, coontails, crabgrass, dog fennel, downybrome, duck salad, fall panicum, fiddleneck, filaree, fireweed, five hook brassia, Flora's paintbrush, Florida pursley, foxtail, goosegrass, groundsel, henbit, jungle rice, knawel, lambsquarters, mustard, naiad, nightshade, pepperweed, pigweed, pineapple weed, plaintain, prickley lettuce, purslane, quackgrass, ragweed, rattail fescue, redmaids, Russian thistle, shepherdspurse, shieldcress, signal grass, silver hair grass, smartweed, Spanish needles, speedwell, submerged aquatic weeds, tansy mustard, wild oats, wiregrass, witchgrass, yellow flower pepperweed.

SODIUM CHLORATE   *Mfg: Drexol, Kerr-McGee and others*

Bermudagrass, bindweed, Canada thistle, hoary cress, johnsongrass, leafy spurge, most annual weeds, paragrass, quackgrass, Russian knapweed.

SULFOMETURON-METHYL, OUST   *Mfg: DuPont*

Annual bluegrass, anise, bahiagrass, barnyard grass, bedstraw, bouncing bet, bromes, buckhorn plaintain, bur clover, Canadian thistle, canary grass, Carolina geranium, chickweed, clovers, curly dock, dandelion, dewberry, dog fennel, doveweed, fescues, fiddleneck, filaree, fleabane, foxtails, goldenrod, groundsel, hemlock, hemp dogbane, hoary cress, honeysuckle, horsetail, Jerusalem artichoke, kochia, kudzu, lambsquarters, little barley, mallow, mayweed, milk thistle, musk thistle, mullein, mustards, ox-eye daisy, pigweed, plantain, poison ivy, prickly lettuce, puncturevine, ragweed, Russian thistle, ryegrass, sow thistle, speedwell, sprangletop, St. John's wort, star thistle, sunflower, sweet clover, tansy mustard, tansy ragwort, turkey mullein, vetch, Virginia pepperweed, wild blackberry, wild carrot, wild oats, yellow nutsedge, yellow rocket.

SULFURIC ACID, N-TAC   *Mfg: Numerous*

Dandilion, dock, goosefoot, groundsel, lambsquarters, London rocket, mustards, nettle, pineapple, shepherdspurse, sorrel, sow thistle.

SUTAN +   *Mfg: Stauffer*

Barnyard grass, bermudagrass seedlings, crabgrass, fall panicum, foxtails, goosegrass, johnsongrass (seedling), nutgrass, sandbur, volunteer sorghum, wild cane.

TCA   *Mfg: Hopkins*

Annual seedling grasses, bermudagrass, bluegrass, crabgrass, foxtails, hilograss, Japanese chess, johnsongrass, paragrass, phragmites, plume grass, quackgrass, torpedograss.

TEBUTHIURON, **GRASLAN, SPIKE**   *Mfg: Elanco*

Alfalfa, alkali sida, annual bluegrass, asters, barley, barnyard grass, bedstraw, bermudagrass, black medic, bluegrass (Kentucky), bouncing bet, bristly oxtongue, bromegrass, broomsedge, buckhorn plaintain, buffelgrass, bull sedge, burclover, buttercup, camphorweed, Canada thistle, Carolina geranium, carpetweed, catsear, cheat, chickweed, chicory, cinquetail, clovers, cocklebur, common mullein, common ragweed, common reed, crabgrass, croton, crowfoot grass, cudweed, dallisgrass, dandelion, dock, dog fennel, fall panicum, fescues, fiddleneck, filaree, five hook bassia, fleabane, foxtail barley, foxtails, giant ragweed, goldenrod, grape, gumweed, henbit, horseweed, itchgrass, Japanese honeysuckle, johnsongrass, knapweed, knotweed, kochia, lambsquarters, little barley, lovegrass, lupine, mallow, milkweed, morningglory, mustard, nightshade, orchardgrass, pepperweed, pigweed, plaintain, poison hemlock, poison ivy, poorjoe, prickly lettuce, prickly sida, puncturevine, purslane, raspberry, reed canary grass, rosering gaillardia, Russian knapweed, Russian thistle, ryegrass, saltbush, sandbur, sedge, shepherdspurse, smartweed, sow thistle, spikeweed, spurge, staghorn sumac, star thistle, strawberry, sunflower, swamp smartweed, telegraph plant, Texas panicum, timothy, trembling aspen, triple awngrass, trumpet creeper, vasey grass, velvetgrass, venue looking glass, vetch, Virginia creeper, Virginia pepperweed, western ragweed, wild carrot, wild oat, wild parsnip, witchgrass, yellow wood-sorrel, and most woody species.

TERBACIL, **SINBAR**   *Mfg: DuPont*

Annual bluegrass, bermudagrass, chickweed, crabgrass, crowfoot, dog fennel, fireweed, Flora's paintbrush, Florida pursley, guinea grass, henbit, horse nettle, johnsongrass (from seed), jungle rice, knotweed, lambsquarters, mustard, natalgrass, nightshade, pigweed, panicum, purslane, ragweed, sandbur, smartweed, sheep sorrel, quackgrass, watergrass, wild geranium, yellow nutgrass.

TERBUTRYN, IGRAN   *Mfg: CIBA-Geigy*

Bedstraw, blue lips, carpetweed, catchweed, chickweed, China lettuce, corn gromwell, corn spurry, cow cockle, false flax, fiddleneck, field pennycress, filaree, foxtails, groundcherry, henbit, Jacobs ladder, jungle rice, knotweed, kochia, lambsquarters, mayweed, meadow foxtail, miners lettuce, mustards, pigweed, pineapple weed, prickly lettuce, ragweed, Russian thistle, shepherdspurse, smartweed, snoweed, speedwell, spring beauty, velvetleaf, watergrass.

THIOBENCARB, BOLERO   *Mfg: Chevron*

Barnyard grass, duck salad, false pimpernel, stem, small flower, spike rush, sprangletop, umbrella sedge, water hyssop.

TRIALLATE, AVADEX BW, FARGO   *Mfg: Monsanto*

Wild oats.

TRICLOPYR, GARLON   *Mfg: Dow Chemical Co.*

Alder, arrowhead, ash, aspen, beech, birch, blackberry, black gum, burdock, cascara, cherry, chicory, choke cherry, cottonwood, crataegus, curled dock, dandelion, dogwood, elderberry, elm, field bindweed, hazel, hickory, hornbean, lambsquarters, locust, maple, mulberry, oaks, persimmon, plaintain, pine, poison oak, poplar, ragweed, salmonberry, sassafras, smartweed, sumac, sweetbay magnolia, sweetgum, Sycamore, thimbleberry, tulip poplar, vetch, wild lettuce, willow, winged elm.

TRIDIPHANE, TAMDEN   *Mfg: Dow*

Barnyardgrass, beggerweed, buffalobur, cocklebur, crabgrass, foxtails, giant ragweed, goosegrass, jimsonweed, ladysthumb, lambsquarters, morningglory, mustard, nightshade, panicum, pigweed, prosomillet, purslane, ragweed, sicklepod, signalgrass, smartweed, starbur, sunflower, velvetleaf, wooly cupgrass.

TRIFLURALIN, TREFLAN   *Mfg: Elanco*

Annual bluegrass, barnyard grass, Brachiaria, bromegrass, carelessweed, carpetweed, cheat, chickweed, crabgrass, Florida pursley, foxtails, goosefoot, goosegrass, johnsongrass, jungle rice, kochia, knotweed, lambsquarters, pigweed, purslane, Russian thistle, sandbur, sprangletop, stinging nettle, stinkgrass, Texas panicum, wild cane, witchweed, wooly cupgrass.

90

VERNOLATE, REWARD, SURPASS, VERNAM

Mfg: Stauffer

Annual morningglory, balloonvine, barnyard grass, carpetweed, cocklebur, coffeeweed, crabgrass, fall panicum, Florida pursley, foxtails, German millet, goosegrass, johnsongrass, lambsquarters, nutgrass, pigweed, purslane, sandbur, sicklepod, velvetleaf, wild cane.

2,4-D

Mfg: Numerous

Alder, alkali mallow, alligator weed, arrowhead, artesmesia, artichoke, asters, Austrian fieldcress, beggarsticks, big sagebrush, bindweed, bitter sneezeweed, bitter watercress, bitter weed, blackeye Susan, bladderwort, blessed thistle, blue lettuce, blue weed, brambles, broomweed, buckbrush, bull thistle, buckrush, burdock, bur ragweed, Canada thistle, carelessweed, carpetweed, catnip, Cherokee rose, chickweed, chicory, coastal sagebrush, cocklebur, coffee weed,, coontail, corn cockle, corn spurry, croton, curly dock, curly indigo, dandelion, day flower, dock, dogbane, duck salad, elderberry, elodea, evening primrose, fanweed, fanwort, fiddleneck, fleabane, flower-of-an-hour, french weed, galinsoga, goatsbeard, goldenrod, gooseweed, ground ivy, gumweed, hazel, heal-all, hemp, henbit, hoary cress, horse nettle, horsetail, Japanese honeysuckle, ironweed, Johnson weed, knotweed, kochia, lambsquarters, leafy spurge, live oak, loco weed, madrone, mallow, manzanita, marestail, marshelder, Mexican weed, milkweed, morningglory, mustard, nettles, oak, orange hawkey, pennycress, pennywort, peppergrass, pepperwood, pigweed, plaintain, poison hemlock, poorjoe, poverty weed, prickly lettuce, primrose, puncturevine, rabbit brush, ragweed, redstem, Russian knapweed, sages, sand sagebrush, St. John's wort, shepherdspurse, slim astor, smartweed, snow-on-the-mountain, snow-on-the-prairie, sow thistle, spike rush, spurge, stinging nettle, sumac, sunflower, tall indigo, tanoak, Tansy ragwort, tan weed, tarweed, tea vine, Texas blueweed, toad flax, turkey pea, umbrella sedge, velvetleaf, vervains, vetch, Virginia creeper, water milfoil, water naiad, wild carrot, winter cress, wild cucumber, wild garlic wild grapes, wild lettuce, wild onion, wild parsnip, wild radish, willow, Yankee weed, yellow rocket, yellow star thistle.

2,4-DB

Mfg: Union Carbide and Vertac

Bull thistle, Canada thistle, cocklebur, coffee weed, curly dock, deadnettle, devil's claw, fanweed, fiddleneck, filaree, goatweed, groundsel, hairy cats ear, hedge bindweed, jimsonweed, knotweed, kochia, lambsquarters, London rocket, morningglory, mustards, nightshade, pigweed, plaintain, prickly lettuce, ragweed, Russian thistle, shepherdspurse, smartweed, stinkweed, sugar beets, sweet clover, tansy mustard, teaweed, velvetleaf, Virginia coffeeleaf, white top, wild beet, turnip, yellow rocket.

NOTES

FUNGICIDES

FUNGICIDES

1. **ACUBA**
   Subdue

2. **AGERATUM**
   Banol
   Banrot
   Bayleton
   Benlate
   Captan
   Daconil-2787
   Subdue
   Termil
   Truban

3. **AGLAONEMA**
   Banrot
   Subdue
   Terrazole
   Truban

4. **AJUGA**
   Banrot
   Benlate
   Captan
   Daconil-2787
   Ornalin
   Rubigan
   Termil

5. **ALDER**
   Benlate
   Captan
   Sulfur

6. **ALMOND**
   (Flowering)
   Bravo
   Daconil-2787
   Ornalin

7. **ALSTROEMERIA**
   Ornalin

8. **ALYSSUM**
   Banol
   Banrot
   Benlate
   Captan
   Ornalin
   Termil
   Terrazole
   Truban

9. **AMELANCHIER**
   Bayleton

10. **ANDROMEDA**
    Subdue

11. **ANTHURIUM**
    Mancozeb
    Ornalin
    Terrazole
    Truban

12. **APHELANDRA**
    Ornalin
    Subdue
    Termil
    Zyban

13. **APPLE**
    (Flowering)
    Benlate
    Captan
    Lime Sulfur
    Streptomycin

14. **ARALIA**
    Banrot

Kocide
Zyban

15. **ARBOR VITAE**
    (Thuja)
    Banrot
    Basic Copper
    Sulfate
    Benlate
    Bordeaux Mixture
    Captan
    Mancozeb
    Subdue

16. **ARCTOSA-**
    PHYLOS
    Truban

17. **ARTEMESIA**
    Ornalin

18. **ASH**
    Bayleton
    Bravo
    Daconil-2787
    Duosan
    Lime Sulfur
    Mancozeb

19. **ASPEN**
    Bayleton

20. **ASTER**
    Banol
    Banrot
    Basic Copper
    Sulfate
    Bayleton
    Benlate
    Captan

95

Copper Oxychloride
Ferbam
Funginex
Karathane
Mancozeb
Maneb
Ornalin
Phaltan
Subdue
Sulfur
Termil
Terrachlor
Topsin-M
Triforine
Truban
Zineb
Ziram
Zyban

21. **AUCUBA**
Mancozeb

22. **AZALEA**
Aliette
Banol
Banrot
Basic Copper
   Sulfate
Bayleton
Benlate
Bordeaux Mixture
Bravo
Captan
Cleary 3336-F
C-O-C-S
Copper Oxychloride
Daconil-2787
Dichlone
Ferbam
Galltrol
Kocide
Mancozeb
Maneb
Nabam

Ornalin
Phaltan
Subdue
Sulfur
Termil
Terrachlor
Thiram
Truban
Zineb
Ziram
Zyban

23. **BABY'S BREATH**
Subdue

24. **BALSAM**
Banrot
Bayleton
Benlate
Captan
Termil
Topsin-M
Truban

25. **BARBERRY**
Bayleton

26. **BEGONIA**
Actidione
Banrot
Basic Copper
   Sulfate
Bayleton
Benlate
Bordeaux
Botran
Captan
Ferbam
Karathane
Kocide
Lime Sulfur
Mancozeb
Maneb
Milban

Ornalin
Rubigan
Subdue
Termil
Truban
Turfcide
Zyban

27. **BELLADONNA**
Benlate
Captan
Karathane

28. **BIRCH**
Bayleton
Lime Sulfur

29. **BLEEDING
HEART**
Ornalin

30. **BOXWOOD**
Aliette
Banrot
Basic Copper
   Sulfate
Bordeaux
Captan
Lime Sulfur
Subdue
Topsin-M
Truban

31. **BRASSIA**
Truban

32. **BROWALLIA**
Banol

33. **BUCKEYE**
Bayleton

34. **BUCKTHORN**
Bayleton

35. **BUFFALO BERRY**
Benlate
Captan
Mancozeb
Maneb

36. **BUTTON BUSH**
Benlate
Captan
Sulfur

37. **CACTUS**
Banrot
Truban

38. **CALADIUM**
Banrot
Benlate
Captan
Daconil-2787
Subdue
Termil
Truban

39. **CALATHEA**
Truban

40. **CALCEOLARIA**
Banrot
Ornalin
Termil

41. **CALENDULA**
Benlate
Captan
Karathane
Rubigan
Sulfur
Terrachlor
Terrazole
Triforine
Truban
Turfcide

42. **CAMELLIA**
Banrot
Bayleton
Benlate
Bordeaux
Captan
Ferbam
Mancozeb
Maneb
Nabam
Ornalin
Terrachlor
Zineb
Ziram
Zyban

43. **CANNA**
Bayleton

44. **CARISSA**
Truban

45. **CARNATIONS**
Banol
Banrot
Basic Copper
    Sulfate
Bayleton
Benlate
Bordeaux
Bravo
Captan
Copper Oxychloride
Daconil-2787
Ferbam
Kocide
Mancozeb
Maneb
Nabam
Ornalin
Phaltan
Plantvax
Subdue
Sulfur

Termil
Terrachlor (PCNB)
Triforine
Truban
Turfcide
Zineb
Ziram
Zyban

46. **CATALPA**
Benlate
Captan
Pipron
Sulfur

47. **CEANOTHUS**
Subdue
Truban

48. **CEDAR**
Actidione
Benlate
Bordeaux
Bravo
Captan
Daconil-2787
Mancozeb
Ornalin
Subdue
Sulfur
Zineb

49. **CELOSIA**
Banol
Banrot
Termil
Terrazole
Topsin-M
Truban

50. **CHERRY
(Flowering)**
Actidione
Ornalin

51. **CHERRY-LAUREL**
    Benlate
    Bravo
    Captan
    Daconil-2787
    Sulfur
    Zyban

52. **CHESTNUT**
    Bayleton
    Benlate
    Captan
    Sulfur

53. **CHOKEBERRY**
    Ornalin

54. **CHRYSANTHE-MUMS**
    Actidione
    Banol
    Banrot
    Basic Copper
        Sulfate
    Bayleton
    Benlate
    Bordeaux
    Botran
    Bravo
    Captan
    Copper Oxychloride
    Daconil-2787
    Ferbam
    Karathane
    Kocide
    Mancozeb
    Maneb
    Milban
    Nabam
    Ornalin
    Phaltan
    Pipron
    Rubigan

Streptomycin
Subdue
Sulfur
Termil
Terrachlor (PCNB)
Terrazole
Topsin-M
Truban
Turfcide
Zineb
Ziram
Zyban

55. **CINERARIA**
    Bayleton
    Benlate
    Captan
    Karathane
    Ornalin
    Termil

56. **CISSUS**
    Zyban

57. **CLEMATIS**
    Galltrol

58. **COLEUS**
    Banol
    Banrot
    Benlate
    Captan
    Ornalin
    Subdue
    Termil
    Truban

59. **COLUMBINE**
    Benlate
    Captan
    Ferbam
    Lime Sulfur
    Ornalin
    Sulfur

60. **CORAL BELLS**
    Ornalin

61. **CORDYLINE**
    Mancozeb

62. **CORNFLOWER**
    Ornalin

63. **COSMOS**
    Benlate
    Captan
    Sulfur

64. **COTONEASTER**
    Kocide
    Subdue

65. **COTTONWOOD**
    Bayleton

66. **CRABAPPLE**
    (Flowering)
    Bayleton
    Benlate
    Bravo
    Captan
    Cleary 3336-F
    Daconil-2787
    Mancozeb
    Sulfur
    Zineb
    Zyban

67. **CRASSULA**
    Bayleton
    Truban

68. **CRAPE MYRTLE**
    Actidione
    Bayleton
    Lime Sulfur
    Ornalin
    Rubigan

Triforine
Zyban

69. **CROSSANDRA**
Ornalin

70. **CROTON**
Banrot

71. **CYCLAMEN**
Benlate
Captan
Ferbam
Ornalin
Termil
Zineb

72. **CYPRESS**
Benlate
Bordeaux
Captan
Mancozeb

73. **DAFFODIL**
Benlate
Captan
Copper Oxychloride
Mertect
Ornalin
Thiram

74. **DAHLIA**
Banol
Banrot
Basic Copper
    Sulfate
Bayleton
Benlate
Bordeaux
Captan
C-O-C-S
Copper Oxychloride
Ferbam
Karathane

Lime Sulfur
Mancozeb
Maneb
Mertect
Pipron
Sulfur
Terrazole
Triforine
Truban
Zineb
Ziram
Zyban

75. **DAISY**
Bayleton
Benlate
Bravo
Captan
Daconil-2787
Ornalin
Subdue
Sulfur

76. **DELPHINIUM**
(Larks Spur)
Banrot
Benlate
Bordeaux
Captan
Daconil-2787
Ferbam
Karathane
Lime Sulfur
Ornalin
Paraformaldehyde
Rubigan
Subdue
Sulfur
Termil
Terrachlor (PCNB)
Zineb
Ziram

77. **DENDROBIUM**
Bayleton

78. **DIANTHUS**
Bayleton
Ornalin
Subdue
Terrazole
Truban

79. **DIEFFEN-
    BACHIA**
Banrot
Benlate
Captan
Ornalin
Streptomycin
Subdue
Terrazole
Truban
Zyban

80. **DOGWOOD**
Basic Copper
    Sulfate
Bayleton
Benlate
Bordeaux
Bravo
Captan
Daconil-2787
Lime Sulfur
Mancozeb
Maneb
Subdue
Sulfur
Zineb
Zyban

81. **DRACAENA**
Banrot
Daconil
Mancozeb
Ornalin
Zyban

82. **DUSTY MILLER**
Banol

83. **ECHEVERIA**
Truban

84. **ELM**
Arbotect
Bayleton
Benlate
Bordeaux
Captan
Lignasan-BLP
Lime Sulfur
Mancozeb
Ornalin
Vapam

85. **EPIPREMNUM**
Banrot
Zyban

86. **EPISCIA**
Ornalin

87. **EUONYMUS**
Actidione
Bayleton
Benlate
Captan
Daconil
Karathane
Kocide
Lime Sulfur
Mancozeb
Ornalin
Rubigan
Triforine
Truban
Zineb

88. **EUPHORBIA**
Ornalin

89. **FATSIA**
Mancozeb
Zyban

90. **FERNS**
Banol
Bayleton
Mancozeb

91. **FESTUCA-FESCUEGRAS**
Benlate
Captan
Daconil-2787
Termil

92. **FICUS**
Banrot
Mancozeb
Ornalin
Truban

93. **FIR**
Banrot
Bayleton
Benlate
Bravo
Busan-72
Captan
Daconil-2787
Mancozeb
Ornalin
Subdue
Truban

94. **FOUR O'CLOCK**
Bayleton

95. **FOXGLOVE**
Benlate
Captan
Sulfur
Truban

96. **FUSCHIA**
Benlate
Captan
Mancozeb
Ornalin
Termil

97. **GARDENIA**
Banrot
Basic Copper
  Sulfate
Bayleton
Benlate
Bordeaux
Captan
Ferbam

98. **GAZANIA**
Ornalin

99. **GERANIUM**
Banol
Banrot
Basic Copper
  Sulfate
Bayleton
Benlate
Bordeaux
Botran
Bravo
Copper Oxychloride
Daconil-2787
Ferbam
Mancozeb
Maneb
Ornalin
Plantvax
Subdue
Termil
Terrazole
Topsin-M
Truban
Zyban

100. GERBERA-TRANSVAAL DAISY
Banrot
Bayleton
Benlate
Captan
Karathane
Ornalin
Truban

101. GLADIOLUS
Arasan
Basic Copper
   Sulfate
Benlate
Bordeaux
Botran
Bravo
Busan-72
Captan
Copper Oxychloride
Daconil-2787
Dyrene
Ferbam
Hexide
Kocide
Mancozeb
Maneb
Mertect
Nabam
Ornalin
Phaltan
Streptomycin
Sulfur
Terrachlor
Thiram
Turfcide
Zineb
Ziram

102. GLOXINIA
Banol
Subdue

103. GOLDEN GLOW
Benlate
Captan
Ferbam
Ziram

104. GYNURA
Banrot
Ornalin
Truban

105. GYPSOPHILIA
Banrot
Galltrol
Ornalin
Truban

106. HAWTHORNE
Actidione
Basic Copper
   Sulfate
Bayleton
Benlate
Bravo
Captan
Daconil-2787
Mancozeb
Sulfur
Zineb
Zyban

107. HEATHER
Subdue

108. HEMLOCK
Truban

109. HIBISCUS (Rose Mallow)
Benlate
Captan
Sulfur

110. HOLLY
Aliette
Banrot
Bayleton
Benlate
Captan
Daconil
Mancozeb
Subdue
Sulfur

111. HOLLYHOCK (Althaea)
Basic Copper
   Sulfate
Bayleton
Benlate
Bordeaux
Bravo
Captan
Copper Oxychloride
Daconil-2787
Ferbam
Lime Sulfur
Sulfur
Zineb
Ziram

112. HONEY SUCKLE
Actidione
Benlate
Captan
Mancozeb
Sulfur

113. HORSECHEST-NUT
Daconil-2787
Mancozeb

114. HOYA
Ornalin
Truban

115. HYACINTH
Benlate
Captan
Mertect
Ornalin
Terrachlor
Thiram
Turfcide

116. HYDRANGEA
Banrot
Benlate
Botran
Captan
Daconil
Ferbam
Karathane
Maneb
Mancozeb
Ornalin
Rubigan
Sulfur
Termil
Zineb

117. IMPATIENS
Banol
Banrot
Ornalin
Subdue
Zyban

118. IRIS
Basic Copper
    Sulfate
Bayleton
Benlate
Bordeaux
Bravo
Busan-72
Chipco 26019
Copper Oxychloride
Daconil-2787
Ferbam

Mancozeb
Maneb
Mertect
Ornalin
Phaltan
Terrachlor
Thiram
Turfcide
Zineb

119. IVY
Banrot
Basic Copper
    Sulfate
Bayleton
Benlate
Bordeaux
Captan
Copper Oxychloride
Ferbam
Kocide
Milban
Ornalin
Subdue
Sulfur
Zineb

120. IXORA
Banrot

**121. JAPANESE-
ANDROMEDA**
Bravo
Daconil-2787

**122. JERUSELUM
THORN**
Triforine

123. JUNIPERS
Aliette
Banrot
Bayleton
Benlate

Bordeaux
Captan
Cleary 3336-F
Ferbam
Lime Sulfur
Mancozeb
Ornalin
Subdue
Sulfur
Truban
Zineb
Zyban

124. KALANCHOE
Banrot
Bayleton
Milban
Ornalin
Truban
Zyban

125. LARKSPUR
Banrot
Karathane
Truban
Turfcide

126. LANTANA
Termil

**127. LAUREL-
SWEET BAY**
Benlate
Bordeaux
Bravo
Captan
Daconil-2787
Ferbam
Mancozeb
Sulfur

**128. LEATHERLEAF
FERN**
Bravo
Daconil-2787

102

129. **LEPTOSPER-**
   MUM
   Ornalin

130. **LEUCOTHE**
   Banrot
   Bayleton

131. **LIATRIS**
   Ornalin

132. **LILAC**
   Basic Copper
      Sulfate
   Bayleton
   Benlate
   Bordeaux
   Captan
   Ferbam
   Karathane
   Lime Sulfur
   Milban
   Nabac
   Phaltan
   Pipron
   Sulfur
   Triforine
   Zineb
   Zyban

133. **LILIES**
   Banol
   Banrot
   Basic Copper
      Sulfate
   Benlate
   Bordeaux
   Bravo
   Captan
   Copper Oxychloride
   Daconil-2787
   Kocide
   Mancozeb
   Maneb

Mertect
Ornalin
Subdue
Termil
Terrachlor
Thiram
Truban
Turfcide
Zineb

134. **LINDEN**
   BASSWOOD
   Benlate
   Bordeaux
   Captan
   Sulfur

135. **LIRIOPE**
   Zyban

136. **LOBELIA**
   Banol
   Banrot
   Benlate
   Captan
   Copper Oxychloride
   Daconil-2787
   Termil

137. **LOCUST**
   Bayleton

138. **LUPINE**
   Ornalin

139. **MAGNOLIA**
   Actidione
   Benlate
   Captan
   Mancozeb
   Terrachlor
   Truban

140. **MAHONIA**
   Bravo
   Daconil-2787

141. **MAPLES**
   Banrot
   Basic Copper
      Sulfate
   Bayleton
   Benlate
   Bordeaux
   Captan
   C-O-C-S
   Duosan
   Mancozeb

142. **MARANTA**
   Banrot
   Daconil
   Truban

143. **MARIGOLD**
   Banol
   Banrot
   Basic Copper
      Sulfate
   Bayleton
   Benlate
   Captan
   Ferbam
   Lime Sulfur
   Mancozeb
   Ornalin
   Ornalin
   Phaltan
   Subdue
   Termil
   Terrazole
   Topsln-M
   Truban
   Ziram

144. **MATRIMONY VINE**
Benlate
Captan
Sulfur

145. **MOCK-ORANGE**
Bayleton

146. **MONARDA**
Ornalin

147. **MONDO-GRASS**
Zyban

148. **MORNING-GLORY**
Benlate
Captan
Ferbam
Ziram

149. **MOUNTAIN LAUREL**
Bayleton

150. **MYRTLE (VINCA)**
Actidione
Benlate
Captan
Lime Sulfur
Subdue
Sulfur
Terrazole
Truban
Zyban

151. **NARCISSUS**
Benlate
Bordeaux
Busan-72
Captan
Copper Oxychloride
Mancozeb
Mertect
Ornalin
Terrachlor
Thiram
Turfcide

152. **NASTURTIUM**
Banrot
Basic Copper
  Sulfate
Benlate
Captan
C-O-C-S
Termil

153. **NAUTILOCALYX**
Banrot

154. **NEPHROLEPIS**
Terrazole

155. **NEPHTHYTIS**
Bayleton

156. **NINEBARK**
Bayleton

157. **NURSERY STOCK (All Kinds)**
Banol
Benlate
Captan
Kathane
Lime Sulfur
Maneb
Terrazole

158. **OAK**
Actidione
Basic Copper
  Sulfate
Bayleton

Benlate
Bordeaux
Captan
C-O-C-S
Daconil-2787
Kocide
Mancozeb
Zineb

159. **ORCHIDS**
Benlate
Captan
Mancozeb
Ornalin
Terrachlor
Truban
Zineb

160. **OREGON GRAPE**
Daconil-2787

161. **ORNAMENTALS (All Kinds)**
Basic Copper
  Sulfate
Benlate
Bordeaux
Captan
Chloropicirin
Cleary 3336-F
Dithane M-45-
  (Fore)
Ferbam
Ficam-W
Funginex
Glyodin
Maneb
Methyl Bromide
Nabac
Thiram
Topsin-M
Vapam
Vorlex
Zineb

162. **OYSTER PLANT**
Bravo
Daconil-2787

163. **PACHYSANDRA-MOUNTAIN SPURGI**
Benlate
Betasan
Captan
Daconil
Dacthal
Kocide
Mancozeb
Maneb
Ornalin
Zineb
Zyban

164. **PALMS**
Banrot
Basic Copper
   Sulfate
Bordeaux
Daconil-2787
Termil
Terramycin

165. **PANSY**
Banol
Banrot
Basic Copper
   Sulfate
Bayleton
Benlate
Bordeaux
Captan
Copper Oxychloride
Mancozeb
Maneb
Subdue
Terrazole
Truban
Zineb
Zyban

166. **PAULOWNIA**
Bayleton

167. **PEAR**
Bayleton

168. **PELARGONIUM**
Ornalin

169. **PELLEA**
Zyban

170. **PEONIES**
Basic Copper
   Sulfate
Benlate
Bordeaux
Captan
Copper Oxychloride
C-O-C-S
Ferbam
Mancozeb
Maneb
Zineb

171. **PEPEROMIA**
Banrot
Ornalin
Subdue
Terrazole
Truban

172. **PERIWINKLE-VINCA**
Benlate
Captan
C-O-C-S
Kocide
Mancozeb
Ornalin
Truban

173. **PETUNIA**
Banol

Banrot
Benlate
Bravo
Captan
Copper Oxychloride
Daconil-2787
Ornalin
Paraformaldehyde
Subdue
Sulfur
Termil
Terrazole
Topsin-M
Truban
Zyban

174. **PHILODENDRON**
Banrot
Benlate
Bravo
Captan
Daconil-2787
Kocide
Mancozeb
Ornalin
Streptomycin
Subdue
Terrazole
Topsin-M
Truban
Zyban

175. **PHLOX**
Actidione
Banrot
Basic Copper
   Sulfate
Bayleton
Benlate
Bordeaux
Captan
C-O-C-S
Ferbam
Nabac

Ornalin
Phaltan
Pipron
Rubigan
Subdue
Sulfur
Triforine
Truban
Ziram
Zyban

176. **PHOTINIA**
Banrot
Bayleton
Daconil
Mancozeb
Rubigan
Subdue
Triforine
Zyban

177. **PIERIS**
Aliette
Daconil-2787
Subdue
Truban

178. **PILEA**
Banrot
Ornalin
Terrazole
Truban

179. **PINES**
Actidione
Aliette
Banrot
Basic Copper
    Sulfate
Bayleton
Benlate
Bordeaux
Bravo
Busan-72

Captan
Citcop
Daconil-2787
Ferbam
Gallex
Lime Sulfur
Mancozeb
Ornalin
Ridomil
Subdue
Thiram
Truban
Ziram

180. **PITTOSPORUM**
Aliette
Banrot
Subdue

181. **PLEOMELE**
Mancozeb

182. **PLUM FLOWER**
Ornalin

183. **PODOCARPUS**
Banrot

184. **POINSETTIAS**
Banol
Banrot
Benlate
Captan
Cleary 3336-F
Copper Oxychloride
Ferbam
Mancozeb
Ornalin
Phaltan
Subdue
Termil
Terrachlor
Topsin-M
Truban

Turfcide
Zyban

185. **POLYSCIAS**
Ornalin

186. **POPLAR**
Bayleton
Benlate
Captan
Daconil
Sulfur
Triforine

187. **POPPY**
Benlate
Captan
Ferbam

188. **PORTULACA**
Banol
Truban

189. **POTENTILLA**
Ornalin

190. **POTHOS**
Subdue
Terrazole
Truban

191. **PRIMROSE**
Benlate
Captan
Daconil-2787
Ferbam
Ornalin
Subdue
Termil

192. **PRIMULA**
Ornalin

193. **PRIVET-
LIGUSTRUM**
Banrot
Bayleton
Benlate
Bordeaux
Bravo
Captan
Daconil-2787

194. **PURPLE LEAF
SAND CHERRY**
Daconil-2787

195. **PYRACANTHA
(Firethorne)**
Bayleton
Benlate
Bravo
Captan
Cleary 336-F
Daconil-2787
Kocide
Mancozeb
Streptomycin
Zyban

196. **QUINCE**
Benlate
Bravo
Captan
Daconil-2787
Zineb

197. **REDWOOD**
Ornalin

198. **RHANNUS**
Truban

199. **RHODODEN-
DRON**
Aliette
Banrot

Basic Copper
   Sulfate
Bayleton
Benlate
Bordeaux
Bravo
Captan
Cleary 3336-F
Copper Oxychloride
Daconil-2787
Mancozeb
Ornalin
Subdue
Sulfur
Terrazole
Truban
Zineb
Zyban

200. **ROSES**
Actidione
Basic Copper
   Sulfate
Bayleton
Benlate
Bordeaux
Botran
Bravo
Captan
Cleary 3336-F
C-O-C-S
Copper Oleate
Copper Oxychloride
Daconil-2787
Dichlone
Ferbam
Funginex
Glyodin
Karathane
Kocide
Lime Sulfur
Mancozeb
Maneb
Milban

Nabam
Ornalin
Parnon
Phaltan
Pipron
Plantvax
Polyram
Rubigan
Streptomycin
Subdue
Sulfur
Termil
Terrachlor
Topsin-M
Triforine
Zineb
Ziram
Zyban

201. **RUSSIAN OLIVE**
Bayleton

202. **SAINTPAULA**
Ornalin

203. **SALVIA**
Banol
Banrot
Bayleton
Benlate
Captan
Ornalin
Subdue
Termil
Terrazole
Topsin-M
Truban

204. **SAXIFRAGA**
Ornalin

205. **SCHEFFLERA**
Aliette
Mancozeb

Ornalin
Subdue

206. **SCILLA**
Ornalin

207. **SCINDAPSUS**
Ornalin

208. **SEDUM**
Bayleton

209. **SENECIO**
Ornalin

210. **SHRUBS**
(All Kinds)
Banol
Benlate
Bordeaux
Captan
Karathane
Lime Sulfur
Thiram

211. **SINNINGIA**
Banrot
Ornalin
Termil
Truban

212. **SNAP DRAGONS**
Banol
Banrot
Basic Copper
   Sulfate
Bayleton
Benlate
Captan
Copper Oxychloride
Ferbam
Karathane
Mancozeb
Maneb

Nabam
Ornalin
Phaltan
Plantvax
Subdue
Sulfur
Termil
Terrachlor
Terrazole
Topsin-M
Triforine
Truban
Turfcide
Zineb
Ziram
Zyban

213. **SPATHI-**
    PHYLLUM
Zyban

214. **SPIREA**
Bayleton
Benlate
Captan
Sulfur

215. **SPRUCE**
Banrot
Benlate
Bordeaux
Bravo
Captan
Daconil-2787
Mancozeb
Ornalin
Subdue
Sulfur

216. **STATICE**
Banrot
Daconil
Mancozeb
Ornalin

217. **STOCKS**
Basic Copper
   Sulfate
Benlate
Captan
Copper Oxychloride
C-O-C-S
Maneb
Paraformaldehyde
Terrachlor
Zyban

218. **STOKESIA**
Truban

219. **STRELITZIA**
Zyban

220. **STROMANTHE**
Truban

221. **SUMAC**
Benlate
Captan
Mancozeb
Maneb
Sulfur

222. **SUNFLOWER**
Bayleton
Benlate
Captan
Sulfur

223. **SWEET PEA**
Basic Copper
   Sulfate
Benlate
Bordeaux
Captan
Copper Oxychloride
Ferbam
Lime Sulfur
Sulfur

Terrachlor
Turfcide

224. **SWEET WILLIAM**
Ornalin

225. **SYCAMORE**
Arbotect
Bayleton
Benlate
Bordeaux
Captan
Daconil
Sulfur
Zineb

226. **SYNGONIUM**
Banrot
Daconil
Mancozeb

227. **TRADESCANTIA**
Banrot

228. **TREES (All Kinds)**
Basic Copper Sulfate
Benlate
Captan
Cleary 3336-F
Karathane
Lime Sulfur
Maneb
Thiram

229. **TULIPS**
Basic Copper Sulfate
Benlate
Bordeaux
Busan-72
Chipco-26019

Copper Oxychloride
Ferbam
Kocide
Mancozeb
Mertect
Ornalin
Terrachlor
Thiram
Turfcide
Zineb

230. **TULIP TREE**
Bordeaux

231. **TURF (General)**
Actidione
Aliette
Apron
Auragreen
Banner
Banol
Basic Copper Sulfate
Bayleton
Benlate
Bravo
Cadminate
Cadminum Chloride
Captan
Chipco-26019
Cleary 3336
Daconil-2787
Demosan
Dyrene
Fore
Fungo
Karathane
Koban
Kromad
Mancozeb
Maneb
Mertect
Nabam

Phaltan
Plantvax
Prevex
Rubigan
Subdue
Terrachlor
Terraneb
Tersan
Topsin-M
Turfcide
Vapam
Vorlex
Zineb

233. **VERBENA**
Banol
Banrot
Benlate
Captan
Ferbam
Rubigan
Subdue
Sulfur
Termil
Terrazole
Topsin-M
Truban
Zyban

234. **VIBURNUM**
Bayleton
Daconil-2787
Mancozeb
Zyban

235. **VINCA**
Banol
Ornalin
Subdue

236. **VIOLETS**
Banol
Basic Copper Sulfate

Bayleton
Benlate
Bordeaux
Captan
Copper Oxychloride
Karathane
Mancozeb
Milban
Ornalin
Subdue
Sulfur
Termil
Terrachlor
Terrazole
Truban
Turfcide

237. **VIRGINIA CREEPER**
Basic Copper
  Sulfate

238. **VITEX**
Bayleton

239. **WALNUT (Black)**
Bayleton
Benlate
Captan
Cyprex
Mancozeb
Zineb

240. **WILLOW**
Bayleton
Benlate
Captan
Lime Sulfur
Sulfur

241. **WOODY ORNAMENTALS (All)**
Banol

242. **YEW (Taxus)**
Benlate
Bordeaux
Captan
Subdue

243. **YUCCA**
Kocide

244. **ZGYGOCACTUS**
Banrot
Subdue
Truban

245. **ZINNIA**
Actidione
Banrot
Basic Copper
  Sulfate
Bayleton
Benlate
Bravo
Captan
Copper Oxychloride
Daconil-2787
Ferbam
Karathane
Lime Sulfur
Mancozeb
Maneb
Milban
Ornalin
Parnon
Phaltan
Pipron

Rubigan
Subdue
Sulfur
Termil
Terrazole
Topsin-M
Triforine
Truban
Zineb
Zyban

FUNGICIDES

ANILAZINE, DYRENE   *Mfg: Mobay*

Alternaria, alternaria leaf spot, Anthracnose, berry spot, black rot, blossom blight, Botrytis fruit rot, Botrytis rot, brown patch, brown spot, cane rust, cane spot, Cercospora leaf spot, copper spot, dollar spot, double spot, downy mildew, early blight, end rot, gray leaf spot, gray mole, gummy stem blight, Helminthosporium spp., late blight, lacy rust, leaf scorch, leaf spot, Phomopsis rot, purple blotch, Rhizoctonia, rust, Sclerotinia dollar spot, Septoria leaf spot, snow mold, twig blight.

BANROT   *Mfg: Mallinckrodt*

Damp off, Fusarium, Phytophthora, Pythium, Rhizoctonia, rootrot, stem rot, Thielaviopsis, water mold

BASIC COPPER CARBONATE   *Mfg: Leffingwell and others*

Blossom blight, brown rot, fire blight, leaf curl, peach blight, peacock spot, Septoia, shot hole, walnut blight.

BASIC COPPER SULFATE, TRI-BASIC   *Mfg: Griffin Co., Phelps Dodge, CP Chemical and others*

Alternaria leaf blight, angular leaf spot, Anthracnose, apple blotch, apple scab, bacterial blight, bacterial canker, bacterial leaf spot, bacterial wilt, bitter rot, black rot blight, blossom rot, blue mold, brown rot, brown rot blossom blight, Cercospora blight, Cercospora leaf spot, citrus scab, downy mildew, early blight, fire blight, greasy spot, gummy stem blight, late blight, leaf blight, leaf curl, leaf mold, leaf spot, melanose, nail head rust, peach blight, peach leaf curl, peacock spot, Phomopsis, pink pitting, powdery mildew, purple blotch, red alga, scab, Septoria leaf spot, Septoria rot, shot hole, smut, Stemphylium leaf spot, walnut blight, white rust, yellow rust.

BENOMYL, BENLATE   *Mfg: DuPont*

Anthracnose, basal rot, bitter rot, black rot, black spot, blackleg, blossom blight, blue mold, Botrytis gray mold, brown patch, brown rot, bunch rot, butter rot, Cercospora foot rot, Cercospora leaf spot, cherry leaf spot, cladosporium leaf mold, crown rot, Diaporthe pod, dyslodia tip blight, dollar spot, downy spot, dry bubble, dutch elm disease, early blight, Eutypa, fly speck, foot rot, fusarium patch, fusarium wilt, Gleosporium, gray mold, greasy spot, green mold, gummy stem blight, late blight, leaf spot, mummy berry, peach scab, penicillium, penicillium rots, Phoma leaf spot, pineapple butt rot, pineapple disease, powdery mildew, rice blast, rusts, scab, Sclerotinia blight, Sclerotinia white mold, sooty blotch, Septoria brown spot, stem blight, stem rot, stem-end rot, strawberry foot rot, stripe smut, target spot, Thielaviopsis, tip blight, white blight, white rot.

BORDEAUX MIXTURE, CUPRIC SULFATE PENTAHYDRATE   *Mfg: Numerous*

Angular leaf spot, Anthracnose, bacterial blast, bacterial blight, bacterial canker, bacterial wilt, bitter rot, black pit, black rot, blossom blight, blotch, brown rot, cane blight, canker gall, Cercospora leaf spot, crown gall, downy mildew, early blight, fire blight, frogeye leaf spot, green rot, gummosis, late blight, late scab, leaf blight, leaf mold, leaf spot, melanose, peach blight, peach leaf curl, peacock spot, Phytophthora, rust, scab, Septoria, shot hole, walnut blight, white mold, wild fire.

CADMINATE   *Mfg: Mallinckrodt*

Brown patch, copper spot, dollar spot, pink patch.

CALCIUM POLYSULFIDE, LIME SULFUR   *Mfg: Miller Chemical & Fert.Corp. and others*

Anthracnose, aphids, apple scab, blister mite, blotch, boxwood canker, Brayobia mite, brown rot, bud mite, cane blight, canker, Coryneum blight, fruit tree leaf roller, mange, maple gall, mealybug, leaf curl, pear leaf, powdery mildew, rust, scab, scales, shot hole, spider mite.

CAPTAFOL, DIFOLATAN   *Mfg: Chevron*

Achlya spp., alternaria, apple scab, Anthracnose, black mold, blossom blight, brown leaf spot, brown rot, Cercospora leaf spot, cherry leaf spot, Coryneum blight, damping-off, downy mildew, early blight, fruit belly rot, fruit rot, gray leaf spot, gummy stem blight, late blight, leaf blight, melanose heart rot, nail head spot, peach leaf curl, pepper blight, Phythium spp., Phytophthora blight, purple blotch, Rhizopus rot, scab, seed rots, shot hole.

CAPTAN   *Mfg: Chevron & Stauffer*

Alternaria, angular leaf spot, Anthracnose, bitter rot, black pox, black rot, blotch, Botryosphaeria (white rot), Botrytis blossom blight, Botrytis flower blight, Botrytis gray mold or berry rot, Brooks fruit spot, brown patch, brown rot, bull's-eye rot, bunch rot, Cercospora spot, copper spot, corn rot and decay, Coryneum blight, damping-off, dead-arm, downy mildew, early blight, flyspeck, frogeye, fruit rots, fusarium, Gleosporium, gray leaf spot, heart rot, Helminthosporium leaf blight, jacket rot, late blight, leaf spot, melanose, melting out, mummy berry, penicillium, petal blight, phoma rot, phomopsis blight, pink rot, purple blotch, Rhizopus, root rots, rust, scab, seedling blights, septoria leaf spot, shot hole, sooty blotch, spur blight, storage rots, transit rots, tuber rot, twig and blossom blight, white mold.

CARBOXIN, VITAVAX   *Mfg: Uniroyal*

Cereal smuts, common bunt, covered smut, damping-off, flag smuts, loose smut, Phythium, Rhizoctonia, seedling disease, seed rot, soreshin, stripe, white mold.

CHLORONEB, DEMOSAN   *Mfg: Kincaid Mfg.*

Damping-off, Phythium spp., Rhizoctonia spp., seedling blight, soreshin, Sclerotium spp.

CHLOROTHALONIL, BRAVO,   *Mfg: Fermenta*
DACONIL 2787

Alternaria, Anthracnose, ascochyta blight, ascochyta ray blight, basal stalk rot, black mold, black spot, blight, blossom blight, Botrytis blight, Botrytis gray mold, Botrytis leaf spot, Botrytis vine rot, brown patch, brown rot blossom blight, brown spot, Cedar apple rust, Cercospora leaf spot, dactylaria leaf spot, dollar spot, downy mildew, early and late blight, fading out, frogeye leaf spot, fruit rot, fusarium, gray leaf spot, gray leaf mold, gray mold, gray snow mold, going out, gummy stem blight, Helminthosporium leaf spot, iris leaf spot, leaf blight, leaf curl, leaf smut, Lophodermium leaf and twig blight, melting out, narrow brown leaf spot, needlecast, ovulinia flower blight, pink rot, Phoma spp., Phythium spp., Phytophthora blight, pod and stem blight, powdery mildew, purple blotch, purple seed stain, red thread, Rhizoctonia blight, Rhizoctonia brown patch, Rhizoctonia fruit rot, Rhizoctonia spp., rice blast, ring spot, rust, scab, scleroderris canker, Sclerotinia dollar spot, scab spot, Septoria leaf spot, sheeth blight, shot hole, shot hole leaf spot, Sirdcocus tip blight, sirrhia brown spot, skin rot, Sphaeropsis leaf spot, stem rust, Swiss needle cast, tan leaf spot, target spot, tip blight, tube rot, web blotch.

CITCOP   *Mfg: Tennessee Chemical*
                                          Co.

Alternaria, bacterial leaf spot, angular leaf spot, bacterial soft rot, bacterial speck, bacterial spot, black rot, bottom rot, brown rot blossom blight, Cercospora leaf spot, Dothistroma, downy mildew, early blight, halo blight, late blight, melanose, needle blights, powdery mildew, purple blotch, red algae, Septoria, Southern leaf blight, walnut blight, web blotch.

COPPER AMMONIUM COMPLEX,   *Mfg: Minerals Research &*
COPPER COUNT-N   *Development*

Alternaria, angular leaf spot, Anthracnose, bacterial blight, bacterial speck, bacterial spot, blight, blossom rot, brown rot, Cercospora leaf spot, downy mildew, early and late blight, greasy spot, halo blight, leaf scorch, leaf spot, melanose, powdery mildew scab, walnut blight.

COPPER HYDROXIDE,   *Mfg: Kocide Chem.,*
BLUE-SHIELD, KOCIDE   *and Others*

Algae, Alternaria blight, angular leaf spot, Anthracnose, bacterial blast, bacterial blight, bacterial leaf blight, bacterial leaf streak, bacterial spot, ball moss, Botrytis blight, brown rot, carrot blight, Cercospora leaf spot, collar rot, common blight, common bunt, Coryneum blight, crown rot, dead bud, downy mildew, early blight, European canker, fire blight, greasy spot, halo blight, Helminthosporium spot blotch, iron spot, late blight, leaf blotch, leaf curl, leaf rust, leaf spot, melanose, peacock spot, phomopsis, pink disease, pink pitting, pseudomonos leaf spot, purple blotch, scab, seed rot, septoria leaf blotch, shot hole, sigatoka, Votutella leaf blight, walnut blight, water mold, xanthomonas leaf spot.

COPPER OXIDE, NORDOX   *Mfg: Nordox, Monterey*
                                          Chemical

Alternaria leaf spot, angular leaf spot, Anthracnose, bacterial blast, bacterial blight, bacterial canker, bacterial speck, bacterial spot, bacterial wilt, berry spot, black leaf spot, black pod rot, black rot, blister blight, blue mold, brown rot, brown rot blossom blight, cane spot, Cercospora leaf spot, damping off, dead bud, downy mildew, early blight, fire blight, frogeye disease, fruit rot, gray leaf spot, greasy spot, gummy stem blight, halo blight, Helminthosporium spot blotch, iron spot, late blight, leaf curl, leaf rust, leaf spot, melanose, peacock spot, pink disease, powdery mildew, purple blotch, scab, septoria leaf blotch, septoria leaf spot, shot hole, sigatoka, walnut blight, white rust, wild fire, yellow rust.

COPPER OXYCHLORIDE, COS   Mfg: J. R. Simplot

Cercospora leaf spot, damping-off, early blight, greasy spot, late blight, leaf curl, leaf spots, melanose.

COPPER OXYCHLORIDE, COCS   Mfg: FMC & Others

Angular leaf spot, Anthracnose, bacterial spot, black rot, brown rot, bud blight, Cercospora leaf spot, cherry leaf spot, Oryneum blight, damping-off, downy mildew, early blight, fire blight, greasy spot, late blight, leaf blight, Macrosporum leaf spot, melanose, peach blight, peach leaf curl, pod spot, powdery mildew, scab, Septoria leaf spot, shot hole, tar leaf spot, twig blight, walnut blight, yellow rust.

CYCLOHEXIMIDE, ACTI-DIONE   Mfg: Nor-Am Ag Products

Brown patch, cherry leaf spot, copper spot, dollar spot, fading out, gray leaf spot, leaf spot, melting out, pink patch, powdery mildew, red thread, rust, snow mold.

DICHLONE, QUINTAR   Mfg: Hopkins Ag Chem.

Algae, Anthracnose, apple scab, black spot, blossom blight, brown rot, cherry leaf spot, Oryneum blight, early blight, fruit rot, gray mold, late blight, peach leaf curl, petal blight, Phoma, twig brown rot.

DICLORAN, BOTRAN   Mfg: Nor-Am Ag Products

Blight, blossom blight, boll rot, Botrytis, Botrytis rot, brown rot, dry rot (Stromatinia), fruit decay, Monilinia, pink rot, Rhizopus, Sclerotinia, Sclerotium, stem canker, white mold, wilt.

DIKAR   Mfg: Rohm & Haas

Apple scab, bitter rot, black rot, brown rot, cedar apple rust, flyspeck, mites, powdery mildew, rusts, sooty blotch.

DINOCAP, KARATHANE   Mfg: Rohm & Haas

Mites, powdery mildew.

DNOC, ELGETOL   Mfg: FMC

Apple scab, brown spot bubble, green mildew, lipstick mold.

DODINE, CYPREX   *Mfg: American Cyanamid*

Anthracnose, bacterial spot, blossom brown rot, downy leaf spot, downy mildew brown leaf spot, leaf blotch, leaf curl, leaf scorch, leaf spot, liver spot, scab.

ETHOXYQUIN, STOP-SCALD   *Mfg: Pennwalt*

Scald.

ETRIDIAZOL, TERRAZOLE,   *Mfg: Uniroyal and*
TRUBAN   *Mallinckrodt*

Damping-off, fusarium, Phytophthora, Phythium, stem rot, Thielaviopsis, stem rot.

FENARIMOL, RUBIGAN   *Mfg: Elanco*

Copper spot, crown rot, dollar spot, fusarium blight, large brown patch, necrotic ring spot, pink and gray snow mold, powdery mildew, red thread, scab, rust, stripe smut, summer patch.

FENTIN HYDROXIDE, DU-TER   *Mfg: Griffin*

Alternaria blight, brown leaf spot, brown spot, Cercospora leaf spot, downy spot, early blight, late blight, leaf blotch, liver spot, pecan scab, powdery mildew, sooty mold.

FERBAM   *Mfg: Rohm & Haas*

Alternaria blight, Anthracnose, apple blotch, bacterial leaf spot, bitter rot, black rot, blossom end rot, blue mold, Botrytis blight, Botrytis rot, Brooks spot, brown leaf spot, brown rot, cedar rust, Cercospora early blight, damping-off, downy mildew, fairy ring, flyspeck, frogeye leaf spot, fruit rot, fruit spot, fusarium rust, fusiform rust, gray mold, leaf blight, leaf spot, mummy berry, peach leaf curl, petal blight, powdery mildew, rust, scab, Septoria late blight, Septoria leaf spot, shot hole, slick spot, sooty blotch, Southern cane rust, spur blight, twig blight.

FOLPET, PHALTAN   *Mfg: Chevron*

Alternaria leaf spot, Anthracnose, bitter rot, black rot, black spot, Botryosphaeria, Brooks fruit spot, brown rot, Cercospora leaf spot, damping-off, dead-arm, Diclymellina leaf spot, downy mildew, early blight, flyspeck, fruit rot, gray mold, greasy spot, late blight, leaf spot, Melanose, powdery mildew, purple blotch, Phythium root rot, Rhizoctonia, root rot, rust, scab, Septoria leaf spot, sooty blotch, storage rot.

FOSETYL-AL, ALIETTE   *Mfg: Rhone Poulenc*

Heart rot, Phytophthora foot rot, Phytophthora root rot, Pythium spp.

FURMECYCLOX, EPIC   *Mfg: Gustafson*

Damping-off, Rhizoctonia spp., sore-kin.

GALLEX   *Mfg: Ag Bio Chem., Inc.*

Burr knot, crown gall, olive knot, rust galls.

GALLTROL-A   *Mfg: Ag Bio Chem., Inc.*

Prevents crown gall.

GLYODIN   *Mfg: Agway, Inc.*

Bitter rot, black rot, black spot, brook spot, brown rot, cherry leaf spot, flyspeck, leaf spot, scab, sooty blotch.

HEXACHLOROPHENE, ISOBAC,   *Mfg: Kalo Labs*
NABAC

Angular leaf spot, bacterial leaf blight, bacterial spot, boll rot, Botrytis spp., damping-off, downy mildew, early and late blight, gray leaf spot, ground rotting, Helminthosporium spp., powdery mildew, Rhizoctonia spp., soreshin, stalk rot.

IMAZALIL, FUNGAFLOR   *Mfg: Pennwalt, FMC,*
                                    Brogdex, Jannsen

Alternaria citri, barley leaf stripe, blue mold, common root rot, diplodia rot, Fusarium spp., green mold, Pennicillum spp., phomopsis stem-end rot, Thielaviopsis.

IPRODIONE, CHIPCO-26019,   *Mfg: Rhone Poulenc*
ROVRAL

Alternaria, Botrytis leaf blight, blossom blight, fruit monilinia brown rot, bottom rot, brown patch, dollar spot, Fusarium blight, Fusarium patch, gray snow mold, Helminthosporium leaf spot, ink spot, lettuce drop, melting out, pink snow mold, purple blotch, red thread, Sclerotinia, shot hole, tulip fire.

LIME SULFUR   *Mfg: Numerous*

Anthracnose, black spot, brown canker, mange, mildew, mites, scab, scale.

MANCOZEB, DITHANE M-45,   *Mfg: Rohm & Haas and*
FORE, MANZATE 200   *DuPont*

Achlya, algae, alternaria, Anthracnose, bacterial spot, bacterial speck, bitter rot, black mold, black rot, black spot, blue mold, Botrytis blight, Botrytis leaf blight, brown rot, bunch rot, bunt, cedar apple rust, Cercospora leaf spot, copper spot, covered kernel smut, covered smut, crown rot, curvularia leaf spot, damping-off, dead-arm, dollar spot, downy mildew, early blight, false loose smut, flyspeck, frogeye leaf spot, fruit rot, Fusarium blight, Fusarium seed decay, Fusarium snow mold, glume blotch, gray leaf mold, gray leaf spot, gummy stem blight, heart rot, Helminthosporium leaf blight, Helminthosporium melting out, Herpobasidium blight, late blight, leaf blight, leaf blotch, leaf mold, leaf spots, Lophodermium fungi, Monochaetia canker, neck rot, pecan scab, petal blight, Phytophthora, purple blotch, purple spot, Phythium blight, red thread, Rhizoctonia, brown patch, rust, scab, seedling blight, seedling rot, Septoria leaf spot, sigatoka, slime mold, sooty blotch, spodixrot, Swiss needle cast, tan spot, Taphrina leaf blister, Votutella blights.

MANEB   *Mfg: DuPont, Rohm &*
                                                 Haas and others

Alternaria leaf spot, angular leaf spot, Anthracnose, bitter rot, black rot, black spot, blast, blossom blight, blue mold, Botrytis leaf blight, brown patch, brown rot, bull's-eye rot, Cercospora early blight, Cercospora leaf spot, curvularia, damping-off, dollar spot, downy mildew, early blight, flyspeck, fruit rot, gray leaf spot, green rot, gummy stem blight, Helminthosporium leaf blight, late blight, leaf spot, melting out, peach leaf curl, Phytophthora, purple blotch, Phythium fruit rot, rust, scab, Septoria late blight, Septoria leaf spot, shot hole, sigatoka disease, sore shin, stemphylium, stem rust, stripe rust, twig blight, white rot, white rust.

METALAXYL, **APRON, RIDOMIL,**   *Mfg: CIBA-Geigy and*
SUBDUE   *Gustafson*

Blue mold, black shank, damping-off, downy mildew, early and late blight, Phytophthora spp., Pythium spp., soot rot, water mold, yellow tuft.

METIRAM, **POLYRAM**   *Mfg: FMC*

Alternaria leaf blight, Anthracnose, asparagus rust, bitter rot, black leg, black rot, black spot, blue mold, brown rot, Botryosphaeria rot, cedar apple rust, Cercospora leaf spot, common scab, damping-off, downy mildew, early blight, European red mite, flyspeck, Fusarium seed piece decay, gray leaf spot, gummy stem blight, Helminthosporium leaf spot, late blight, pecan scab, scab, sooty blotch.

MILBAN   *Mfg: Mallinckrodt*

Powdery mildew.

MP-11   *Mfg: Monsanto*

Mold inhibition.

MYCO-SHIELD   *Mfg: Pfizer*

YTD in citrus, bacterial spot, fire blight, ice nucleating bacteria.

NABAM and ZINC, **DITHANE A-40,**   *Mfg: Rohm & Haas*
DITHANE D-14

Alternaria, Anthracnose, black rot, blast, Botrytis blight, Botrytis leaf blight, brown rot, Cercospora leaf spot, Cladosporium gray mold, curvularia disease, downy mildew, early blight, Fusarium decay, gummy stem, Helminthosporium leaf blight, late blight, leaf spot, petal blight, purple blotch, ripe rot, rust, Septoria leaf spot, stemphylium, white rust.

OPP   *Mfg: Dow and Pennwalt*

Post harvest molds.

OXYCARBOXIN, **PLANTVAX**   *Mfg: Uniroyal*

Rusts.

PCNB, **TERRACHLOR, TURFCIDE**   *Mfg: Uniroyal*

Botrytis storage rot, bottom rot, brown patch, brown rot, bulb dry rot, bulb rot, bunt, camellia flower blight, club root, common smut, crown rot, damping-off, leaf drop, leaf spot, Magnolia leaf spot, neck rot, petal blight, Rhizoctonia, root rot, Sclerotinia, snow mold, Southern blight, stem rot, white mold, white rot, wire stem.

PIPERALIN, **PIPRON**   *Mfg: Elanco*

Powdery mildew.

PROPAMOCARB-HYDRO-   *Mfg: Nor-Am*
CHLORIDE, **BANOL, PREVEX**

Cottony blight, damping-off, grease spot, Pythium blight, Pythium spp., Phytophthora spp., rootrot, rusts.

PROPICONAZOL, BANNER, TILT   *Mfg: CIBA-Geigy*

Anthracnose, brown patch, dollar spot, powdery mildew, red thread, rust, Selenophoma stem eyespot, stripe smut.

SODIUM ARSENITE   *Mfg: LA Chemical and others*

Black measles, dead-arm, phomopsis shoot necrosis.

SOPP   *Mfg: Dow and Pennwalt*

Post harvest decay organisms.

STREPTOMYCIN   *Mfg: Pfizer & Merck*

Bacterial blight, bacterial spot, bacterial wilt, black leg, blue mold, fire blight, soft rot, wildfire.

SULFUR   *Mfg: Numerous*

Almond mite, Atlantic mite, blister mite, brown rot, bud mite, cherry leaf spot, citrus leaf mite, citrus rust mite, European red mite, flyspeck, leaf spot, Pacific mite, peach scab, powdery mildew, red spider mite, russet mite, rust, scab, Septoria leaf spot, silver mite, six-spotted mite, sooty blotch, spider mite, thrips, two-spotted mite, Vienna spider mite.

TCMTB, BUSAN-30, BUSAN-72   *Mfg: Buckman Labs, Inc.*

Covered smut, Curvularia spp., damping-off, fusarium spp., Helmin-thosporium spp., loose smut, Penicillium spp., Phythium, rhizoctonia, seed decay, seedling blight, seedling rots, smut, soreshin, stinking smut.

TERRAMYCIN   *Mfg: Pfizer*

Bacterial spot, fire blight, lethal decline, lethal yellowing disease, pear decline, X-disease.

THIABENDAZOLE, ARBORTECT, MERTECT   *Mfg: Merck*

Anthracnose, basal rot, black rot, blue mold, blue mold rot, Botrytis, brown patch, brown spot, bull's-eye rot, Cercospora foot rot, Cercospora leaf spot, cobweb, crown rot, dutch elm disease, dollar spot, dry bubble, foot rot, frogeye leaf spot, fusarium patch, glume blotch, green mold, gray mold, narrow brown leaf spot, neck rot, pod blight, purple seed stain, rice blast, Sclerotinia rot, scurf, sheath blight, stem, blight, stem end rot, stem rot, tuber rot, wet bubble, white mold.

THIOPHANATE, CLEARY 3336,   *Mfg: W. A. Cleary*
TOPSIN-E

Anthracnose, black spot, Botrytis flower blight, brown patch, bulb rot, copper spot, crown rot, dollar spot, fusarium, Helminthosporium, leaf blight, leaf spot, ovulinia blight, Phythium blight, red thread, root rot, rust, scab, skin blight, spot-kleen, stem rot, stripe smut.

THIOPHANATE-METHYL, FUNGO,   *Mfg: Mallinckrodt,*
TOPSIN-M   *Pennwalt*

Anthracnose, bitter rot, black knot, black rot, blossom blight, Botrytis fruit rot, Botrytis gray mold, Botrytis spp., brown leaf spot, brown patch, brown rot, ceratocystis, Cercospora leaf spot, copper spot, coscomyces, dendrophoma leaf blight, diplocarpon leaf scorch, dollar spot, early and late blight, eye spot, flyspeck, foot rot, frogeye leaf scorch, dollar spot, early and late blight, eye spot, flyspeck, foot rot, frogeye leaf spot, fruit brown rot, Fusarium blight, Fusarium patch, Gloessporium spp., gray mold, late blight, leaf blight, leaf scorch, leaf spot, monilinia, peach scab, Penicillium spp., pod and stem blight, powdery mildew, purple seed stain, red thread, rhizoctonia spp., scab, sooty blotch, strawbreaker, stripe smut, Thielaviopsis spp., white rot.

THIRAM, ARASAN   *Mfg: DuPont and*
                                         Rohm and Haas
                                         and others

Alternaria spp., apple blotch, banana fruit spots, basal rot, bitter rot, black pox, black rot, Botryosphaeria, Brooks spot, brown patch, brown rot, cedar apple rust, Cercospora early blight, crown rust, damping-off, dollar spot, early blight, flyspeck, fruit rot, gray mold, leaf blight, leaf scorch, leaf spot, Rhizopus rot, root rot, scab, scruff, seed decay, smut, surface mold of bananas, snow mold, sooty blotch, stem rot.

TRIADIMEFON, BAYLETON   *Mfg: Mobay*

Anthracnose, barley scald, brown patch, copper spot, dollar spot, flower blight, Fusarium blight, Fusarium patch, gray snow mold, leaf blight, leaf blotch, petal blight, pine rust, powdery mildew, pink snow mold, red thread, rusts, Sclerotinia spp., snow mold, strip smut, tip blight.

TRIADIMENOL, BAYTAN   *Mfg: Mobay and Gustafson*

Flag smut, leaf rust, loose smut, powdery mildew, Septoria, stem rust, stinking smut, stripe rust, take-all.

TRICYCLAZOLE, BEAM   *Mfg: Elanco*

Rice blast.

TRIFORINE, FUNGINEX   *Mfg: EM Labs*

Anthracnose, black spot, blossom blight, brown rot, leaf spot, powdery mildew, mummy berry, rust, scab.

TRUBAN   *Mfg: Mallinckrodt*

Damping-off, Phytophthora spp., Pythium spp.

VAPAM   *Mfg: Stauffer*

Club root, dutch elm disease, Fusarium, many weeds, nematodes, oak root fungus, Phytophthora, Phythium, rhizoctonia, Sclerotinia, soil insects, Verticillium.

VINCLOZOLIN, ORNALIN, RONILAN   *Mfg: BASF*

Bortrytis blight, Bortrytis rot, brown rot, ciborinia, gray mold, molinia, ovulinia, Sclerotinia spp., Stromatinia.

ZINC COPOSIL, CUPRIC ZINC SULFATE COMPLEX   *Mfg: Numerous*

Anthracnose, brown rot, bunch rot, Cercospora leaf spot, downy mildew, early and late blights, fire blight, leaf curl, leaf spot, powdery mildew, scab, Septoria, shot hole, walnut blight.

ZINEB   *Mfg: Rohm & Haas*

Alternaria leaf blight, Alternaria spp., Anthracnose, apple blotch, bitter rot, black rot, blue mold, Botrytis blight, Botryosphaeria fruit rot, Brooks spot, cedar rust, Cercospora early blight, Cercospora leaf spot, cobweb, Dactylium spp., damping-off, downy mildew, early blight, fading-out, fire blight, flyspeck, frogeye leaf spot, fruit rot, fruit russet, Fusarium seed decay, Fusarium seed piece decay, Fusarium spp., gray leaf spot, greasy spot, green mold Helminthosporium blight, late blight, leaf rust, leaf spot, mildew, Mycogone (bubbles), Mycogone spp., peach leaf curl, pecan scab, petal blight, Phomopsis spp., Phytophthora, Phythium blight, quince rust, ripe rot, rust, scab, Septoria late blight, Septoria leaf spot, shot hole, slime mold, sooty blotch, soreshin, stem rust, Southwestern cotton rust, Trichoderma spp., Verticillium spp., white rust.

ZIRAM
Mfg: Rohm & Haas, Pennwalt and Others

Anthracnose, black rot, blossom blight, Botrytis spp., brown rot, bull's-eye rot, Cercospora early blight, evergreen blackberry canker, evergreen canker, fruit brown rot, fruit rot, Fusiform rust, gray mold, leaf spot, mummy berry, peach leaf curl, pecan scab, petal blight, powdery mildew, rust, scab, Septoria late blight, shot hole, twig blight.

ZYBAN
Mfg: Mallinckrodt

Anthracnose, black spot, downy mildew, flower blight, leaf blight, powdery mildew, rust, scab stem and twig blight.

NOTES

GROWTH
REGULATORS

GROWTH REGULATORS

1. **ABELIA**
Atrinal
Maintain CF-125

2. **ACACIA**
Atrinal
Embark
Maintain CF-125
MH-30

3. **ACANTHOPANAX**
Hormodin

4. **AESCHYNANTHUS**
Atrinal

5. **AFRICAN VIOLET**
Hormodin

6. **AGERATUM**
A-Rest
Gibberellic Acid
Hormodin

7. **AGLAOMENA**
Gibberellic Acid

8. **AJUGA**
Gibberellic Acid

9. **ALDER**
Maintain CF-125
MII-30

10. **ALTERNANTHERA**
A-Rest

11. **ALYSSUM**
Atrinal

12. **ANDROMEDA**
Hormodin

13. **APPLE**
Florel
Hormodin
Tre-hold

14. **ARBOR VITAE**
(Thuja)
Atrinal
Hormodin
Rootone

15. **ARBUTUS**
Hormodin

16. **ARDISIA**
Hormodin

17. **ASH**
Atrinal
Maintain CF-125

18. **ASPEN**
Maintain

19. **ASTER**
B-Nine
Gibberellic Acid

20. **AZALEA**
Atrinal
B-Nine
Cycocel
Florel
Gibberellic Acid
Hormodin
Off-Shoot-O
Rootone-F

21. **BALSAM**
A-Rest

22. **BARBERRY**
Atrinal
Hormodin
Rootone

23. **BAYBERRY**
Atrinal
Hormodin

24. **BEAUTY BERRY**
Hormodin

25. **BEAUTY BUSH**
Hormodin

26. **BEECH**
Hormodin

27. **BEGONIA**
Atrinal
Cycocel
Hormodin
Rootone-F

28. **BIRCH**
Hormodin

29. **BITTER SWEET**
Hormodin

30. **BLACKBERRY**
Hormodin

31. **BLEEDING**
HEART
B-Nine

32. **BLUEBEARD**
Hormodin

33. **BLUE BELL**
A-Rest

34. **BLUEBERRY**
Hormodin

35. **BOTTLEBRUSH**
Atrinal
Embark

36. **BOUGAIN-**
VILLEA
Atrinal
Hormodin

37. **BOXWOOD**
Hormodin
Rootone-F

38. **BRAZILIAN**
PEPPER TREE
Embark

39. **BROMELIADS**
Florel

40. **BROOM**
Hormodin

41. **BUCKTHORN**
Atrinal

42. **BUTTERFLY**
BUSH
Atrinal
Hormodin

43. **CACTUS**
Hormodin

44. **CALCEOLARIA**
Cycocel

45. **CALIANDRA**
Maintain CF-125

46. **CAMELLIA**
Hormodin

47. **CANDY TUFT**
Hormodin

48. **CAPEWEED**
Embark

49. **CARNATIONS**
Accel
Hormodin
Rootone-F

50. **CAROB**
Embark
Florel

51. **CATALPA**
Hormodin
NAA
Stik

52. **CEDAR**
Maintain-A

53. **CHASTE-TREE**
Hormodin

54. **CHERRY**
Hormodin
Maintain-A

55. **CHERRY-**
LAUREL
Atrinal

56. **CHESTNUT**
Hormodin

57. **CHINESE ELM**
Atrinal

58. **CHOKE-BERRY**
Hormodin

59. **CHRYSANTHE-**
MUMS
A-Rest
B-Nine
Gibberellic Acid
Hormodin
Rootone-F

60. **CINERARIA**
B-Nine

61. **CINQUEFOIL**
Hormodin

62. **CISSUS**
Atrinal
MH-30

63. **CITRUS**
Tre-Hold

64. **CLEMATIS**
Hormodin

65. **CLERODEN-**
DRUM
A-Rest
Atrinal
Hormodin

66. **CLEYERA**
Atrinal

67. **CLOCKVINE**
Hormodin

68. **COLEUS**
Gibberellic Acid

69. **COSMOS**
B-Nine

70. **COTONEASTER**
Atrinal
Hormodin
MH-30
Off-Shoot-O

71. **COTTONWOOD**
Atrinal
Maintain CF-125

72. **CRABAPPLE**
Florel
Hormodin
Tre-hold

73. **CRASSULA**
Gibberellic Acid
Hormodin

74. **CREEPER**
Hormodin

75. **CREEPING LIPPIA**
Atrinal

76. **CREPE MYRTLE**
Atrinal
Hormodin
Maintain CF-125

77. **CROTON**
Hormodin

78. **CRYPTOMERIA**
Hormodin

79. **CURRANT**
Hormodin

80. **CYCLAMEN**
Gibberellic Acid

81. **CYPRESS**
Atrinal

82. **DAFFODIL**
Florel

83. **DAHLIA**
A-Rest
Gibberellic Acid
Hormodin

84. **DAISY**
Embark
Gibberellic Acid

85. **DAPHNE**
Hormodin

86. **DELPIIINIUM (Larks spur)**
Gibberellic Acid

87. **DEUTZIA**
Hormodin

88. **DEW BERRY**
Hormodin

89. **DOGWOOD**
Hormodin
Rootone

90. **DOVETREE**
Hormodin

91. **DRACAENA**
A-Rest
Gibberellic Acid
Hormodin

92. **DUTCH-MANSPIPE**
Hormodin

93. **EASTER LILY**
A-Rest

94. **ELAENAGNUS**
Atrinal
Hormodin
Maintain CF-125

95. **ELDER**
Hormodin
Maintain-A

96. **ELM**
Atrinal
Hormodin
Maintain-A
Maintain CF-125
MH-30
NAA
Stik
Tre-hold

97. **ESCALLONIA**
Hormodin

98. **EUCALYPTUS-GUM**
Atrinal
MH-30
Maintain CF-125

99. **EUGENIA-SURINAM CHERRY PITANGA**
Atrinal
Maintain CF-125
MH-30

100. **EUONYMUS**
Atrinal
Maintain CF-125

101. **FATSHEDERA**
Atrinal
Gibberellic Acid

102. **FICUS**
Atrinal
Embark

103. **FIR**
Maintain CF-125

104. **FONTANESIA**
Hormodin

105. **FORSYTHIA-GOLDEN BELL**
Atrinal
Gibberellic Acid
Hormodin
MH-30

106. **FOXGLOVE**
Gibberellic Acid

107. **FRANKLINIA**
Hormodin

108. **FRINGE TREE**
Hormodin

109. **FUSCHIA**
Atrinal
Gibberellic Acid
Hormodin

110. **GARDENIA**
Atrinal
B-Nine
Hormodin

111. **GAZANIA**
Atrinal

112. **GELSEMIUM**
Atrinal

113. **GERANIUM**
A-Rest
Cycocel
Florel
Gibberellic Acid
Hormodin
Rootone-F

114. **GERBERA**
Gibberellic Acid

115. **GERMANDER**
Gibberellic Acid
Hormodin

116. **GLOXINIA**
B-Nine
Gibberellic Acid

117. **GOLDEN CHAIN**
Hormodin

118. **GRAPE**
Hormodin

119. **GRAPTO-PETALUN**
Gibberellic Acid

120. **HACKBERRY**
Atrinal

121. **HARDY-ORANGE**
Atrinal

122. **HAWTHORN**
Atrinal
Hormodin

123. **HAZELNUT**
Hormodin

124. **HEATH**
Hormodin

125. **HEATHER**
Hormodin

126. **HEMLOCK**
Hormodin

127. **HIBISCUS**
Embark
Hormodin
Maintain CF-125

128. **HICKORY**
Maintain-A

129. **HOLLY**
Atrinal
Embark
Gibberellic Acid
Hormodin
Rootone

130. **HONEYSUCKLE**
Atrinal
Hormodin
MH-30

131. **HORSECHEST-NUT**
NAA
Stik

132. **HYACINTH**
A-Rest
Florel

133. **HYDRANGEA**
B-Nine
Gibberellic Acid
Hormodin

134. **ICE PLANT**
Embark
Maintain CF-125
MH-30

135. **IVY**
Atrinal
Embark
Gibberellic Acid
Maintain CF-125
MH-30

136. **JASMINE**
Atrinal
Hormodin
Maintain CF-125

137. **JERUSALEM CHERRY**
Gibberellic Acid

138. **JETBEAD**
Hormodin

139. **JUNIPERS**
Atrinal
Hormodin
Maintain CF-125
Off-Shoot-O
Rootone

140. **KALANCHOE**
A-Rest
Atrinal
B-Nine

141. **KERRIA**
Hormodin

142. **KNOTWEED**
Hormodin

143. **KUDZU**
Maintain CF-125

144. **LABURNO-CYTISUS**
Hormodin

145. **LANTANA**
Atrinal
Hormodin

146. **LARKSPUR**
Gibberellic Acid

147. **LAUREL**
Hormodin

148. **LAUREL FIG**
Atrinal
Maintain CF-125

149. **LAVENDER**
Hormodin

150. **LEUCOTHOE**
Hormodin

151. **LILAC**
Hormodin
Rootone

152. **LILY SCALES**
Hormodin

153. **LINDEN**
Hormodin

154. **LIPPIA**
Atrinal

155. **LOCUST**
Hormodin
MH-30
Maintain-A

156. **LONDON PLANE TREE**
Atrinal

157. **MAGNOLIA**
Hormodin

158. **MAINDENHAIR-TREE**
Hormodin

159. **MANZANITA**
Hormodin

160. **MAPLES**
Atrinal
Gibberellic Acid
Hormodin
Maintain-A
Maintain CF-125
MH-30
NAA
Stik
Tre-hold

161. **MARANTA**
Gibberellic Acid

162. **MARIGOLD**
B-Nine

163. **MATRIMONY-VINE**
Hormodin

164. **MELALEUCA**
Embark
Maintain CF-125

165. **MELASTROMA**
Hormodin

166. **MOCK-ORANGE**
Hormodin

167. **MULBERRY**
Atrinal
Hormodin

168. **MYRTUS**
MH-30

169. **NARCISSUS**
Florel

170. **NEPHTHYTIS**
A-Rest

171. **NINE BARK**
Hormodin

172. **OAK**
Atrinal
Gibberellic Acid
Hormodin
MH-30
Maintain-A
Tre-hold

173. **OLEANDER**
Atrinal
Embark
Hormodin
Maintain CF-125
MH-30

174. **OLIVES**
Atrinal
Embark
Florel
Hormodin
Maintain-A
NAA
Tre-hold

175. **ORANGE-JESSAMINE**
Atrinal

176. **ORIKA**
Hormodin

177. **ORNAMENTALS (General)**
IBA (Hormodin)
NAA (Rootone)
Pinolene

178. **OSAGE ORANGE**
Hormodin

179. **OSMANTHUS**
Atrinal
Hormodin

180. **OXALIS**
Gibberellic Acid

181. **PACHYSANDRA-MOUNTAIN SPURGE**
Hormodin
Rootone

182. **PANSY**
Gibberellic Acid

183. **PATHOS**
A-Rest

184. **PEAR**
Tre-hold

185. **PEA SHRUB**
Hormodin

186. **PELTHANTHRUS**
Gibberellic Acid

187. **PENSTEMON**
Hormodin

188. **PEPEROMIA**
Gibberellic Acid

189. **PEPPER TREE**
Maintain CF-125

190. **PERIWINKLE**
Atrinal
Hormodin

191. **PETUNIA**
A-Rest
B-Nine
Gibberellic Acid
Hormodin

192. **PHILODENDRON**
A-Rest
Hormodin

193. **PHLOX**
Hormodin

194. **PHOTINIA**
Atrinal
Embark
Hormodin

195. **PILEA**
A-Rest

196. **PINES**
Hormodin
Maintain CF-125
Pro-Shear

197. **PITTOSPORUM**
Atrinal
Embark
MH-30

198. **PLUMBAGO**
Maintain CF-125

199. **PODOCARPUS**
Atrinal
Embark

200. **POINSETTIAS**
A-Rest
B-Nine
Cycocel
Gibberellic Acid
Hormodin

201. **PONCIRUS**
Hormodin

202. **POPLAR**
Gibberellic Acid
Hormodin
Maintain A
MH-30

203. **PREMISES**
Maintain CF-125
MH-30
Short-Stop

204. **PRIVET-
LIGUSTRUM**
Atrinal
Embark
Hormodin
Maintain CF-125
MH-30
Off-Shoot-O

205. **PURPLE
PASSION**
A-Rest

206. **PYRACANTHA**
Atrinal
Embark
Hormodin
MH-30
Rootone-F

207. **QUINCE**
Hormodin

208. **RAPHIOLEPIS**
Atrinal
Embark

209. **REDWOOD**
Maintain CF-125

210. **RETINOSPORA**
Hormodin

211. **RHAMNUS-
TALL HEDGE**
Off-Shoot-O

212. **RHODODEN-
DRON**
Gibberellic Acid
Hormodin
Rootone

213. **RHOEO**
Gibberellic Acid

214. **ROSES**
Accel
Atrinal
Gibberellic Acid
Hormodin
Rootone

215. **SAGE**
Hormodin

216. **SALVIA**
A-Rest
B Nine

217. **SCHEFFLERA**
A-Rest
Atrinal

218. **SEDUM**
Gibberellic Acid

219. **SEQUOIA**
Hormodin

220. **SHRIMP PLANT**
Atrinal

221. **SHRUBS
(General)**
MH-30

222. **SILVER BELL**
Hormodin

223. **SNAKE-PLANT**
Hormodin

224. **SNAP DRAGON**
Hormodin

225. **SNOW BELL**
Hormodin

226. **SNOW BERRY**
Hormodin

227. **SOURWOOD**
Hormodin

228. **SPEEDWELL**
Hormodin

229. **SPIREA**
Hormodin

230. **SPRINGSCENT**
Hormodin

231. **SPRUCE**
Hormodin
Maintain CF-125

232. **STATICE**
Gibberellic Acid

233. **STEVIA**
Hormodin

234. **STOCKS**
Gibberellic Acid

235. **ST. JOHNSWORT**
Hormodin

236. **SWEET GUM**
MH-30

237. **SWEET LEAF**
Hormodin

238. **SYCAMORE**
Atrinal
MH-30
Maintain-A
Tre-hold

239. **TALLHEDGE BUCKTHORN**
Florel

240. **TAMARACK-JAPANESE LARCH**
MH-30

241. **TECOMARIA**
Atrinal

242. **TOLMIA**
Gibberellic Acid

243. **TREES (General)**
MH-30
Tre-hold

244. **TRUMPET-CREEPER**
Hormodin

245. **TULIPS**
A-Rest

246. **TULIP TREE**
Hormodin

247. **TURF (General)**
Cutlass
Embark
Gibberellic Acid
Limit
Maintain CF-125
MH-30
MH-30T
PO-San
Short Stop
TGR-Poa Control

248. **UMBRELLA-PINE**
Hormodin

249. **VERBENA**
Atrinal
Hormodin

250. **VIBURNUM**
Atrinal
Embark
Hormodin
MH-30
Rootone

251. **VIOLETS**
Gibberellic Acid

252. **WALNUT (Black)**
MH-30
Maintain-A

253. **WANDERING JEW**
Gibberellic Acid

254. **WAXMYRTLE**
Hormodin

255. **WEIGELIA**
Hormodin

256. **WILLOW**
Atrinal
Hormodin
Maintain-A
Maintain CF-125

257. **WINTERGREEN**
Hormodin

258. **WISTERIA**
Hormodin

259. **WITCH-HAZEL**
Hormodin

260. **XYLOSMA**
Atrinal
Embark
Maintain CF-125
MH-30

261. **YELLOW WOOD**
Hormodin

262. **YEW (Taxus)**
Hormodin
Off-shoot-O
Rootone-F

263. **YUCCA**
Gibberellic Acid

264. **ZEBRINA**
Gibberellic Acid

265. **ZELKOVA**
Hormodin

266. **ZINNIA**
B-Nine

GROWTH REGULATORS

A-REST   *Mfg: Elanco Prod. Co.*

Controls plant height.

ACCEL   *Mfg: Abbot Labs*

Increase lateral branching, increases flower production.

ATRINAL, ATRIMMEC   Mfg: Maag Agrochemicals
                                              PBI/Gordon

Chemical pinching agent, growth retardation for landscape plantings of ground
covers, hedges and shrubs and to suppress flowering of fruit formulation of
certain ornamentals.

B-NINE   *Mfg: Uniroyal*

Chemical height retardant and promotes flowering. Also reduces leaf injury to
certain plants caused by air pollution.

BONZI   *Mfg: Zoecon*

Controls plant height.

CLIPPER   *Mfg: ICI Americas*

Controls plant height.

CUTLASS   *Mfg: Elanco*

Used for the growth reduction of turfgrass species.

CYCOCEL   *Mfg: American Cyanamid*

Controls plant height.

EMBARK   *Mfg: PBI/Gordon*

Reduces plant growth, suppresses seedhead formations, reduces mowings,
inhibits fruiting of some species.

FLOREL, ETHREL   *Mfg: Union Carbide*

Induces flowering of ornamentals, prevents fruit set of certain ornamentals,
defoliates certain plants, and increases lateral branching.

GIBBERELLIC ACID   *Mfg: Merck, Abbott*

Elongate growth of plants, causes earlier flowering, causes more profuse flowering, causes larger flowers, and on Bermuda turf to maintain growth and prevent color change during periods of cold stress or light frost.

HORMODIN-(IBA)   *Mfg: Merck & Co.*

Promotes rooting of ornamental cuttings.

IBA   *Mfg: Numerous*

Promotes rooting of plant cuttings.

LIMIT   *Mfg: Monsanto*

Used on turf to reduce vegetative growth and to suppress seedhead formation.

MH-30 (GRO-SLO)   *Mfg: Uniroyal and Others*

To reduce plant growth.

MAINTAIN, MAINTAIN CF-125   *Mfg: Leffingwell*
and MAINTAIN-A

Controls plant height and a growth sprout retardant, controls broadleaf weeds in turf.

NAA   *Mfg: Union Carbide,*
                              Amvac, FMC

Promotes rooting of plant cuttings, prevents seed pod and nut formation and prevents olive set.

OFF-SHOOT-O   *Mfg: Procter and Gamble*

Used to chemically pinch plants.

PINOLENE, WILT-PRUF   *Mfg: Miller Chemical Co.*
OR VAPORGUARD   *and Nursery*
                                             Specialty Products

Retards water transpirations and maintains healthy foliage.

PO-SAN   *Mfg: Mallinckrodt*

Post-emergence control of poa annual to prevent seed head formation.

PRO-SHEAR   *Mfg: Abbott*

Increase lateral bud set and branch development.

ROOTONE-F   *Mfg: Abbott Labs*

Promotes rooting.

SHORT-STOP   *Mfg: Stauffer*

Used to suppress tall fescue seed head formation.

STIK   *Mfg: FMC Corp.*

To prevent seed pod and nut formation.

TGR POA ANNUA CONTROL   *Mfg: O. M. Scott*

Slows the growth of Poa annua in active growing turf.

TRE-HOLD   *Mfg: Union Carbide*

Controls new growth of tree sprouts after trimming.

NOTES

GLOSSARY

SPRAYER CALIBRATION

Calibration of your spraying equipment is very important. It should be done at least every other day of operation to insure application of the proper dosages. This is probably the most important step in your whole spraying operation since applying incorrect amounts may do much more damage than good.

If a lower rate is desired it may be obtained by increasing the speed, reducing the speed, increasing the pressure or changing to a larger nozzle or a combination of the three.

GENERAL CALIBRATION

I. Method I.
    A. Measure out 660 feet.
    B. Determine the amount of spray put out in traveling this distance at the desired speed.
    C. Use this formula:

$$\text{gallons/acre} = \frac{\text{gallons used in 660 feet} \times 66}{\text{swath width in feet}}$$

   *D. Fill the tank with the desired concentration.

II. Method II.
    A. Fill spray tank and spray a specified number of feet.
    B. After spraying refill tank measuring the quantity of material needed for refilling.
    C. Use this formula:

$$\text{gallons/acre} = \frac{43560 \times \text{gallons delivered}}{\text{swath length (ft.)} \times \text{swath width (ft.)}}$$

   *D. Fill the tank with the desired concentrate.

III. Method III.
    A. Measure 163 feet in the field.
    B. Time tractor in 163 feet. Make two passes to check accuracy.
    C. At edge of field adjust pressure valve until you catch 2 pints (32 ounces) of spray in the same amount of time it took to run the 163 feet. Be sure tractor is at the same throttle setting. You are now applying 20 gallons per acre on a 20 inch nozzle boom spacing.
    D. For each inch of nozzle spacing on boom, increase time by 5% or reduce the volume by 5%.

---

*If you have calibrated your rig and it is putting out 37 gpa, the required dosage is 4 lbs. actual/acre. Therefore, for every 37 gallons of carrier (water, oil, etc.) in the spray tank you add 4 lbs. of active material.

USEFUL FORMULAE

1. To determine the amount of active ingredient needed to mix in the spray tank.

   No. of gallons or pounds =

   $$\frac{\text{No. of acres to be sprayed} \times \text{pounds active ingredient required per A}}{\text{pounds active ingredient per gallon or per pound}}$$

2. To determine the amount of pesticide needed to mix a spray containing a certain percentage of the active ingredient.

   No. of gallons or pounds =

   $$\frac{\text{gallons of spray desired} \times \% \text{ active ingredient wanted} \times 8.345}{\text{lbs. active ingredient per gallon or pound} \times 100}$$

3. To determine the percent active ingredient in a spray mixture.

   Percent =

   $$\frac{\text{lbs. or gallons of concentrate used (not just active ingredient)} \times \% \text{ active ingredient in the concentrate}}{\text{gallons of spray} \times 8.345 \text{ (weight of water/gallon)}}$$

4. To determine the amount of pesticide needed to mix a dust with a given percent active ingredient.

   pounds material =

   $$\frac{\% \text{ active ingredient wanted} \times \text{lbs. of mixed dust wanted}}{\% \text{ active ingredient in pesticide used}}$$

5. To determine the size of pump needed to apply a given number of gallons/acre.

   pump capacity =

   $$\frac{\text{gallons/acre desired} \times \text{boom width (feet)} \times \text{mph}}{495}$$

6. To determine the nozzle capacity in gallons per minute at a given rate/acre and miles/hour.

   Nozzle capacity =

   $$\frac{\text{gallons/acre} \times \text{nozzle spacing (inches)} \times \text{mph}}{5940}$$

7. To determine the acres per hours sprayed.

   Acres per hour =

   $$\frac{\text{swath width (inches)} \times \text{mph}}{100}$$

8. To determine the rate of speed in miles per hour.
   1. Measure off a distance of 300 to 500 feet.
   2. Measure in seconds the time it takes the tractor to go the marked off distance.

3. Multiply .682 times the distance traveled in feet and divide product by the number of seconds.

$$MPH = \frac{.682 \times \text{distance}}{\text{seconds}}$$

9. To determine the nozzle flow rate.
Time the seconds necessary to fill a pint jar from a nozzle.
Divide the number of seconds into 7.5.

$$\text{gallons/minute/nozzle} = \frac{7.5}{\text{seconds}}$$

10. To determine the gallons per minute per boom.
Figure out the gallons/minute/nozzle and multiply by the number of nozzles.

11. To determine the gallons per acre delivered.

$$\frac{5940 \times \text{gallons/minute/nozzle}}{\text{nozzle spacing (inches)} \times \text{mph}} = \text{gpa}$$

12. To determine the acreage sprayed per hour.

$$\text{acres sprayed/hour} = \frac{\text{boom width (feet)} \times \text{mph}}{12}$$

This allows 30% of time for filling, turning, etc.

13. Sprayer Tank Capacity
Calculate as follows:
1. Cylindrical Tanks:
Multiply the length in inches times the square of the diameter in inches and multiply the product by .0034.
length × diameter squared × .0034 = number of gallons.
2. Elliptical Tanks:
Multiply the length in inches times the short diameter in inches times the long diameter in inches times .0034.
length × short diameter × long diameter × .0034 = number of gallons.
3. Rectangular Tanks:
Multiply the length times the width times the depth in inches and multiply the product by .004329.
length × width × depth × .004329 = number of gallons.

14. To determine the acres in a given area.
Multiply the length in feet times the width in feet times 23. Move the decimal point 6 places to the left to give the actual acres.

CONVERSION TABLES (U.S.)

Linear Measure -
>1 foot - 12 inches
>1 yard - 3 feet
>1 rod - 5.5 yards — 16.5 feet
>1 mile - 320 rods — 1760 yards — 5280 feet

Square Measure -
>1 square foot (sq. ft.) — 144 square inches (sq. inch)
>1 square yard (sq. yd.) — 9 sq. feet
>1 square rod (sq. rd.) — 272.25 sq. ft. — 30.25 sq. yd.
>1 acre - 43560 sq. ft. — 4840 sq. yds. — 160 sq. rds.
>1 square mile - 640 acres

Cubic Measure -
>1 cubic foot (cu. ft.) — 1728 cubic inches (cu. in.) 29.922
>liquid quarts = 7.48 gallons
>1 cubic yard - 27 cubic feet

Liquid Capacity Measure -
>1 tablespoon - 3 teaspoons
>1 fluid ounce - 2 tablespoons
>1 cup - 8 fluid ounces
>1 pint - 2 cups — 16 fluid ounces
>1 quart - 2 pints — 32 fluid ounces
>1 gallon - 4 quarts — 8 pints — 128 fluid ounces

Weight Measure -
>1 pound (lb.) - 16 ounces
>1 hundred weight (cwt.) - 100 pounds
>1 ton - 20 cwt. — 2000 pounds

Rates of Application -
>1 ounce/sq. ft. — 2722.5 lbs./acre
>1 ounce/sq. yd. — 302.5 lbs./acre
>1 ounce/100 sq. ft. — 27.2 lbs./acre
>1 pound/100 sq. ft. — 435.6 lbs./acre
>1 pound/1000 sq. ft. — 43.6 lbs./acre
>1 gallon/acre — 3 ounce/1000 sq. ft.
>5 gallons/acre — 1 pint/1000 sq. ft.
>100 gallons/acre — 2.5 gallons/1000 sq. ft. — 1 quart/100 sq. ft.
>100 lbs./acre — 2.5 lbs./1000 sq. ft.

Important Facts -

Volume of sphere — diameter^3 × .5236

Diameter — circum. × .31831

Area of circle — dia. ^{2} × .7854

Area of ellipse — prod. of both dia. × .7854

Vol. of cone — area of base × 1/3 ht.

1 cu. ft. water = 7.5 gallons = 62.5 lbs.

Pressure in psi — ht. (ft.) × .434

1 acre = 209 feet square

ppm = % × 10,000

% — ppm
  10,000

1% by volume — 10,000 ppm

TABLE OF CONVERSION FACTORS

To Convert From	To	Multiply By
Cubic feet	gallons	7.48
Cubic feet	liters	28.3
Gallons	milliliters	3785
Grams	pounds	.0022
Grams/liter	parts/million	1000
Grams/liter	pounds/gallon	.00834
Liters	cubic feet	.0353
Milligrams/liter	parts/million	1
Milliliters/gallons	gallons	.0026
Ounces	grams	28.35
Parts/million	grams/liter	.001
Parts/million	pounds/million gallons	8.34
Pounds	grams	453.59
Pounds/gallon	grams/liter	111.83

1 gram = .035 ounce   1 metric ton = 1000 kg. = 2,205 lbs.

1 kilogram = 2.2 lbs.   1 hectare = 2.5 acres

1 quintal = 100 kg. = 221 lbs. 1 meter = 39.4 inches

           1 kilometer = .6 mile

CONVERSION TABLE

1 kilogram (kg) = 1000 grams (g) = 2.2 lbs.
1 gram (g) = 1000 milligrams (mg) = .035 ounce
1 liter = 1000 milliliters (ml) or cubic centimeters (cc) = 1.058 quarts
1 milliliter or cubic centimeter = .034 fluid ounce
1 milliliter or cubic centimeter of water weighs 1 gram
1 liter of water weighs 1 kilogram
1 lb. = 453.6 grams
1 ounce = 28.35 grams
1 pt. of water weighs approximately 1 lb.
1 gallon of water weighs approximately 8.34 lbs.

1 gallon = 4 qts. = 3.785 liters
1 qt. = 2 pts. = .946 liters
1 pt. = .473 liters
1 fluid ounce = 29.6 milliliters or cubic centimeters

1 part per million (ppm) = 1 milligram/liter
                         = 1 milligram/kilogram
                         = .0001 percent
                         = .013 ounces by in 100 gallons of water
1 percent = 10.000 ppm
          = 10 grams per liter
          = 10 grams per kilogram
          = 1.33 ounces by weight per gallon of water
          = 8.34 pounds/100 gallons of water

.1 percent = 1000 ppm = 1000 milligrams/liter
.01 percent = 100 ppm = 100 milligrams/liter
.001 percent = 10 ppm = 10 milligrams/liter
.0001 percent = 1 ppm = 1 milligram/liter

CHEMICAL ELEMENTS

Name	Symbol	Atomic Weight	Valance
Aluminum	Al	26.97	3
Antimony	Sb	121.76	3, 5
Arsenic	As	74.91	3, 5
Barium	Ba	137.36	2
Bismuth	Bi	209.00	3, 5
Boron	B	10.82	3, 0
Bromine	Br	79.916	1, 3, 5, 7
Cadmium	Cd	112.41	2
Calcium	Ca	40.08	2
Carbon	C	12.01	2, 4
Chlorine	Cl	35.457	1, 3, 5, 7
Cobalt	Co	58.94	2, 3
Copper	Cu	63.57	1, 2
Fluorine	F	19.00	1
Hydrogen	H	1.008	1
Iodine	I	126.92	1, 3, 5, 7
Iron	Fe	55.85	2, 3
Lead	Pb	207.21	2, 4
Magnesium	Mg	24.32	2
Mercury	Hg	200.61	1, 2
Molybdenum	Mo	95.95	3, 4, 6
Nickel	Ni	58.69	2, 3
Nitrogen	N	14.008	3, 5
Oxygen	O	16.00	2
Phosphorus	P	30.98	3, 5
Potassium	K	39.096	1
Selenium	Se	78.96	3
Silicon	Si	28.06	4
Silver	Ag	107.88	1
Sodium	Na	22.997	1
Sulfur	S	32.06	2, 4, 6
Thallium	Tl	204.29	1, 3
Tin	Sn	118.70	2, 4
Titanium	Ti	47.90	3, 4
Uranium	U	238.17	4, 6
Zinc	Zn	65.38	2

WIDTH OF AREA COVERED TO ACRES PER MILE TRAVELED

Width of Strip (feet)	Acres/mile
6	.72
10	1.21
12	1.45
16	1.93
18	2.18
20	2.42
25	3.02
30	3.63
50	6.04
75	9.06
100	12.1
150	18.14
200	24.2
300	36.3

TEMPERATURE CONVERSION TABLE RELATIONSHIP OF CENTIGRADE AND FAHRENHEIT SCALES

°C	°F	°C	°F
-40	-40	25	77
-35	-31	30	86
-30	-22	35	95
-25	-13	40	104
-20	-4	45	113
-15	5	50	122
-10	14	55	131
-5	23	60	140
0	32	80	176
5	41	100	212
10	50		
15	59		
20	68		

PROPORTIONATE AMOUNTS OF DRY MATERIALS

Water	Quantity of Material				
100 gallons	1 lb.	2 lbs.	3 lbs.	4 lbs.	5 lbs.
50 gallons	8 oz.	1 lb.	24 oz.	2 lbs.	2 1/2 lbs.
5 gallons	3 tbs.	1 1/2 oz.	2 1/2 oz.	3 1/4 oz.	4 oz.
1 gallon	2 tsp.	3 tsp.	1 1/2 tbs.	2 tbs.	3 tbs.

PROPORTIONATE AMOUNTS OF LIQUID MATERIALS

Water	Quantity of Material		
100 gallons	1 qt.	1 pt.	1 cup
50 gallons	1 pt.	1 cup	1/2 cup
5 gallons	3 tbs.	5 tsp.	2 1/2 tsp.
1 gallon	2 tsp.	1 tsp.	1/2 tsp.

MILES PER HOUR CONVERTED TO FEET PER MINUTE

MPH	fpm
1	88
2	176
3	264
4	352

EMULSIFIABLE CONCENTRATE PERCENT RATINGS
IN POUNDS ACTUAL PER GALLON

%EC	lbs./Gallon
10-12	1
15-20	1.5
25	2
40-50	4
60-65	6
70-75	8
80-100	10

CONVERSION TABLE FOR LIQUID FORMULATIONS

Concentration of Active Ingredient in Formulations, lbs./gal.

Rate Desired Lbs./A	1	2	2.5	3	4	5	6
			(ml of formulation per 100 square feet)				
1	8.67	4.33	3.47	2.89	2.17	1.73	1.44
2	17.3	8.67	6.93	5.78	4.33	3.47	2.89
3	26.0	13.0	10.4	8.67	6.50	5.20	4.33
4	34.8	17.4	13.9	11.6	8.69	6.95	5.79
5	43.4	21.7	17.4	14.5	10.0z	8.68	7.24
6	52.1	26.0	20.8	17.4	13.0	10.4	8.68
7	60.8	30.4	24.3	20.3	15.2	12.2	10.1
8	69.4	34.7	27.8	23.1	17.4	13.9	11.6
9	78.1	39.0	31.2	26.0	19.5	15.6	13.0
10	86.7	43.3	34.7	28.9	21.7	17.3	14.4

*Example: To put out a 100 sq. ft. plot at the rate of 5 lbs./A active ingredient using a formulation containing 4 lbs./gal. active ingred., use 10.9 ml. of the 4 lbs./gal. form. & distribute evenly.

CONVERSION TABLE FOR DRY FORMULATIONS

Concentration of Active Ingredient in Formulation

Rate Desired Lbs./A	100%	90%	80%	75%	70%	60%	50%	40%	30%	25%	20%	10%	5%
				(Grams of formulation per 100 square feet)								
1	1.04	1.16	1.30	1.39	1.49	1.74	2.08	2.60	3.47	4.17	5.21	10.4	20.8
2	2.08	2.31	2.60	2.78	2.98	3.47	4.17	5.21	6.94	8.33	10.4	20.8	41.7
3	3.12	3.47	3.90	4.17	4.46	5.20	6.25	7.81	10.4	12.5	15.6	31.2	62.5
4	4.17	4.63	5.21	5.55	5.95	6.94	8.33*	10.4	13.9	16.7	20.8	41.7	83.3
5	5.21	5.79	6.51	6.94	7.44	8.68	10.4	13.0	17.4	20.8	26.0	52.1	104
6	6.25	6.94	7.81	8.33	8.93	10.4	12.5	15.6	20.8	25.0	31.2	62.5	125
7	7.29	8.10	9.11	9.72	10.4	12.1	14.6	18.2	24.3	29.2	36.4	72.9	146
8	8.33	9.26	10.4	11.1	11.9	13.9	16.7	20.8	27.8	33.3	41.7	83.3	167
9	9.37	10.4	11.7	12.5	13.4	15.6	18.7	23.4	31.2	37.5	46.9	93.7	187
10	10.4	11.6	13.0	13.9	14.9	17.4	20.8	26.0	34.7	41.7	52.1	104	208

*Example: To put out a 100 sq. ft. plot at the rate of 4 lbs./A active ingredient using a formulation containing 50% active ingredient, use 8.33 grams of the 50% formulation and distribute evenly over the 100 sq. ft.

151

CONVERSION TABLE FOR GRANULAR FORMULATIONS

Concentration of Active Ingredient in Formulation

Rate Desired Lbs./A	20%	15%	10%	7.5%	5%	4%	3%	2%	1%
				(Grams of formulation per 100 square feet)					
1	5.2	6.94	10.4	13.86	20.8	26.0	34.66	52.0	104.0
2	10.4	13.9	20.8	27.7	41.7	52.0	69.3	104.0	208.0
3	15.6	20.8	31.2	41.6	62.5	78.0	103.9	156.0	312.0
4	20.8	27.8	41.7	55.4	83.3	104.0	138.6	208.0	416.0
5	26.0	34.7	52.1	69.3	104.0*	130.0	173.3	260.0	520.0
6	31.2	41.6	62.5	83.2	125.0	156.0	207.9	312.0	624.0
7	36.4	45.6	72.9	97.0	146.0	182.0	242.6	364.0	728.0
8	41.7	55.5	83.3	110.9	167.0	208.0	277.3	416.0	832.0
9	46.9	62.5	93.7	124.7	187.0	234.0	311.9	468.0	936.0
10	52.1	69.4	104.0	138.6	208.0	260.0	346.6	520.0	1040.0
15	78.0	104.1	156.0	207.9	312.0	390.0	519.2	780.0	1560.0
20	104.0	138.8	208.0	277.2	416.0	520.0	693.2	1040.0	2080.0
25	130.0	173.5	260.0	346.5	520.0	650.0	866.5	1300.0	2600.0
30	156.0	208.2	312.0	415.8	624.0	780.0	1039.8	1560.0	3120.0

*Example: To put out a 100 sq. ft. plot at the rate of 5 lbs./A active ingredient using a formulation containing 5% active material, use 104.0 grams of the 5% formulation and distribute it evenly over the 100 sq. ft.

DETERMINE THE NUMBER OF ROWS TO THE ACRE

Length of Rows

Rows/Acre	32"	36"	38"	40"	60"
1	16335	14520	13756	13068	8712
2	8168	7260	6878	6534	4356
3	5445	4840	4585	4356	2904
4	4084	3630	3439	3267	2178
5	3267	2904	2751	2614	1742
6	2723	2420	2293	2178	1452
7	2334	2074	1965	1867	1245
8	2042	1815	1719	1634	1089
9	1815	1613	1528	1452	968
10	1634	1452	1376	1307	871
11	1485	1320	1251	1188	792
12	1361	1210	1156	1089	726
13	1257	1117	1058*	1005	670
14	1167	1037	982	933	622
15	1089	968	917	871	581
16	1021	908	760	817	545
17	961	854	809	769	512
18	908	807	764	726	484
19	860	764	724	688	459
20	817	726	688	653	436
21	778	691	655	622	415
22	743	660	625	594	396
23	710	631	598	568	379
24	681	605	573	544	363
25	653	581	550	523	348
26	628	558	529	503	335
27	605	538	509	484	323
28	583	519	491	467	311
29	563	501	474	450	300
30	545	484	459	436	290

*Example: A grower's field is 1058 long furrowed out on 38-inch centers. Therefore, every 13 rows across the field represents an acre.

QUICK CONVERSIONS

TEMPERATURE		LENGTH		VOLUME	
°C	°F	cm	inch	liters	quarts
100	212	2½	1	1	1.1
90	194	5	2	2	2.1
80	176	10	4	3	3.2
70	158	20	8	4	4.2
60	140	30	12	5	5.3
50	122	40	16	6	6.3
40	104	50	20	7	7.4
35	95	60	24	8	8.5
30	86	70	28	9	9.5
25	77	80	32		
20	68	90	36		
15	59	100	39		
10	50	200	79		
5	41		feet		
0	32	300	10		
-5	23	400	13		
-10	14	500	16		
-15	5	1,000	33		
-20	-4				
-25	-13				
-30	-22				
-40	-40				

QUICK CONVERSIONS

kg./ha.		lb./A
1	....	0.9
2	....	1.8
3	....	2.7
4	....	3.6
5	....	4.5
10	....	9
20	....	18
30	....	27
40	....	36
50	....	45
60	....	54
70	....	62
80	....	71
90	....	80
100	....	89
200	....	180
300	....	270
400	....	360
500	....	450
600	....	540
700	....	620
800	....	710
900	....	800
1000	....	890
2000	....	1800

kg./ha.		ton/A
3000	....	1¼
4000	....	1¾
5000	....	2¼
6000	....	2¾
7000	....	3
8000	....	3½
9000	....	4
10000	....	4½
11000	....	5
12000	....	5½
13000	....	5¾
14000	....	6¼
15000	....	6¾
16000	....	7
17000	....	7½
18000	....	8
19000	....	8½
20000	....	9

USEFUL MEASUREMENTS

LENGTH
   1 mile = 80 chains = 8 furlongs = 1760 yards = 5280 ft. = 1.6 kilometers
   1 chain = 22 yards = 4 rods, poles or perches = 100 links

WEIGHT
   1 long ton = 20 cwt. = 2240 lbs.
   1 lb. = 16 ozs. = 454 grams = 0.454 kilograms
   1 short ton = 2000 lbs.
   1 metric ton = 2204 lbs. = 1000 kilograms

AREA
   1 acre = 10 sq. chains = 4840 sq. yards = 43560 sq. ft. = 0.405 hectares
   1 sq. mile = 640 acres = 2.59 kilometers
   1 hectare = 2.471 acres

VOLUME
   1 gal. = 4 quarts = 8 pints = 128 fluid ozs. = 3.785 liters
   1 fluid oz. = 2 tablespoons = 4 dessertspoons = 8 teaspoons = 28.4 c.c's
   1 liter = 1000 c.c.'s = 0.22 Imperial gallon = 1.76 pints

CAPACITIES
   Cylinder-diameter ^{2}/ × depth × 0.785 = cubic feet
   Rectangle-breadth × depth × length = cubic feet
   Cubic feet × 6.25 = gallons

QUICK CONVERSIONS

1 pint/acre	= 1 fluid oz./242 sq. yards
1 gal./acre	= 1 pint/605 sq. yards
1 lb./acre	= 1 oz./300 sq. yards
1 cwt./acre	= 0.37 oz./sq. yard
1 m.p.h.	= 88 ft./minute
3 m.p.h.	= 1 chain/15 sec.
1 liter/hectare	= 0.089 gal./acre
1 kilogram/hectare	= 0.892 lb./acre
1 c.c./100 liters	= 0.16 fl. oz./100 gallons
125 c.c./100 liters	= 1 pint/100 gallons
1 gm./100 liters	= 0.16 oz./100 gallons

A strip 3 ft. wide × 220 chains ⎞
A strip 4 ft. wide × 165 chains ⎬   1 acre
A strip 5 ft. wide × 132 chains ⎠

CONVERSION FACTORS USED IN CALCULATION

Convert	To	By

gram (gm.)   kilogram (kg.)   move decimal 3 places to left
Example: 2000 gm. = 2.0 kg.

gram (gm.)   milligram (mg.)   move decimal 3 places to right
Example: 2.0 gm. = 2000 mg.

gram (gm.)   pound (lb.)   divide by 454
Example: 658 gm./ ÷ 454 = 1.45 lb.

gram/pound   percent (%)   divide by 4.54
Example: 90 gm./lb. ÷ 4.54 = 19.8%

gram/ton   percent   multiply by 11,
Example: 45 gm./ton × 11 = 495 = .00495% move decimal 5 places to left

kilogram (kg.)   gram   move decimal 3 places to right
Example: 5.5 kg. = 5500 gm.

milligram (mg.)   gram   move decimal 3 places to left
Example: 95 mg. = 0.095 gm.

percent   gram/pound   multiply by 4.54
Example: 25 × 4.54 = 113.5 gm./lb.

percent   parts/million (ppm.) move decimal 4 places to right
Example: .025% = 250 ppm.

percent   gram/ton   divide by 11,
Example: .011 ÷ 11 = .001 = 100 gm./ton move decimal 5 places to right

pound   gram   multiply by 454
Example: 0.5 lb. × 454 = 227 gm.

ppm   percent   move decimal 4 places to left
Example: 100 ppm = 0.01%

GRAMS/GALLONS TABLE

Gallons PPM	5	10	15	20	25	50	75	100	150	200	300	400
5	0.1	0.2	0.3	0.4	0.5	1.0	1.4	1.9	2.8	3.8	5.7	7.6
10	0.2	0.4	0.6	0.8	1.0	1.9	2.8	3.8	5.7	7.6	11.0	15.0
15	0.3	0.6	0.9	1.1	1.4	2.8	4.3	5.7	8.5	11.0	17.0	23.0
20	0.4	0.8	1.1	1.5	1.9	3.8	5.7	7.6	11.0	15.0	23.0	30.0
25	0.5	0.9	1.4	1.9	2.4	4.7	7.1	9.5	14.0	19.0	28.0	38.0
50	0.9	1.9	2.8	3.8	4.7	9.5	14.0	19.0	28.0	38.0	57.0	76.0
75	1.4	2.8	4.3	5.7	7.1	14.0	21.0	28.0	43.0	57.0	85.0	114.0
100	1.9	3.8	5.7	7.6	9.5	19.0	28.0	38.0	57.0	76.0	114.0	151.0
125	2.4	4.7	7.1	9.5	12.0	24.0	36.0	47.0	71.0	95.0	142.0	189.0
150	2.8	5.7	8.5	11.0	14.0	28.0	43.0	57.0	85.0	114.0	170.0	227.0
175	3.3	6.6	9.9	13.0	17.0	33.0	50.0	66.0	99.0	133.0	199.0	265.0
200	3.8	7.6	11.0	15.0	19.0	38.0	57.0	76.0	114.0	151.0	227.0	303.0
250	4.7	9.5	14.0	19.0	24.0	47.0	71.0	95.0	142.0	189.0	284.0	379.0
300	5.7	11.0	17.0	23.0	28.0	57.0	85.0	114.0	170.0	227.0	341.0	454.0
400	7.6	15.0	23.0	30.0	38.0	76.0	114.0	151.0	227.0	303.0	454.0	606.0

STANDARD MEASUREMENTS

Measure of Length (Linear Measure)

$$
\begin{array}{rcl}
4 \text{ inches} &=& 1 \text{ hand} \\
9 \text{ inches} &=& 1 \text{ span} \\
12 \text{ inches} &=& 1 \text{ foot} \\
3 \text{ feet} &=& 1 \text{ yard} \\
6 \text{ feet} &=& 1 \text{ fathom} \\
5\frac{1}{2} \text{ yards - } 16\frac{1}{2} \text{ feet} &=& 1 \text{ rod} \\
40 \text{ poles} &=& 1 \text{ furlong} \\
8 \text{ furlongs} &=& 1 \text{ mile} \\
5,280 \text{ feet } = 1,760 \text{ yards} &=& 320 \text{ rods } = 1 \text{ mile} \\
3 \text{ miles} &=& 1 \text{ league}
\end{array}
$$

Measure of Surface (area)

$$
\begin{array}{rcl}
144 \text{ square inches} &=& 1 \text{ square foot} \\
9 \text{ square feet} &=& 1 \text{ square yard} \\
30\frac{1}{4} \text{ square yards} &=& 1 \text{ square rod} \\
160 \text{ square rods} &=& 1 \text{ acre} \\
43,560 \text{ square feet} &=& 1 \text{ acre} \\
640 \text{ square acres} &=& 1 \text{ square mile} \\
36 \text{ square miles} &=& 1 \text{ township}
\end{array}
$$

Surveyor's Measure

$$
\begin{array}{rcl}
7.92 \text{ inches} &=& 1 \text{ link} \\
25 \text{ links} &=& 1 \text{ rod} \\
4 \text{ rods} &=& 1 \text{ chain} \\
10 \text{ square chains} &=& 160 \text{ square rods } = 1 \text{ acre} \\
640 \text{ acres} &=& 1 \text{ square mile} \\
80 \text{ chains} &=& 1 \text{ mile} \\
1 \text{ Gunter's chain} &=& 66 \text{ feet}
\end{array}
$$

Metric Length

$$
\begin{array}{rcl}
1 \text{ inch} &=& 2.54 \text{ centimeters} \\
1 \text{ foot} &=& .305 \text{ meter} \\
1 \text{ yard} &=& .914 \text{ meter} \\
1 \text{ mile} &=& 1.609 \text{ kilometers} \\
1 \text{ fathom} &=& 6 \text{ feet} \\
1 \text{ knot} &=& 6,086 \text{ feet} \\
3 \text{ knots} &=& 1 \text{ league} \\
1 \text{ centimeter} &=& .394 \text{ inch} \\
1 \text{ meter} &=& 3.281 \text{ feet} \\
1 \text{ meter} &=& 1.094 \text{ yards} \\
1 \text{ kilometer} &=& .621 \text{ mile}
\end{array}
$$

Troy Weight

$$24 \text{ grains} = 1 \text{ pennyweight}$$
$$20 \text{ pennyweight} = 1 \text{ ounce}$$
$$12 \text{ ounces} = 1 \text{ pound}$$

Liquid Measure

$$2 \text{ cups} = 1 \text{ pint}$$
$$4 \text{ gills} = 1 \text{ pint}$$
$$16 \text{ fluid ounces} = 1 \text{ pint}$$
$$2 \text{ pints} = 1 \text{ quart}$$
$$4 \text{ quarts} = 1 \text{ gallon}$$
$$31\frac{1}{2} \text{ gallons} = 1 \text{ barrel}$$
$$2 \text{ barrels} = 1 \text{ hogshead}$$
$$1 \text{ gallon} = 231 \text{ cubic inches}$$
$$1 \text{ cubic foot} = 7.48 \text{ gallons}$$
$$1 \text{ teaspoon} = .17 \text{ fluid ounces (1/6 oz.)}$$
$$3 \text{ teaspoons (level)} = 1 \text{ tablespoon (1/2 oz.)}$$
$$2 \text{ tablespoons} = 1 \text{ fluid ounce}$$
$$1 \text{ cup (liquid)} = 16 \text{ tablespoons (8 oz.)}$$
$$1 \text{ teaspoon} = 5 \text{ to } 6 \text{ cubic centimeters}$$
$$1 \text{ tablespoon} = 15 \text{ to } 16 \text{ cubic centimeters}$$
$$1 \text{ fluid ounce} = 29.57 \text{ cubic centimeters}$$

Cubic Measure (Volume)

$$1{,}728 \text{ cubic inches} = 1 \text{ cubic foot}$$
$$27 \text{ cubic feet} = 1 \text{ cubic yard}$$
$$2{,}150.42 \text{ cubic inches} = 1 \text{ standard bushel}$$
$$231 \text{ cubic inches} = 1 \text{ standard gallon (liquid)}$$
$$1 \text{ cubic foot} = 4/5 \text{ of a bushel}$$
$$128 \text{ cubic feet} = 1 \text{ cord (wood)}$$
$$7.48 \text{ gallons} = 1 \text{ cubic foot}$$
$$1 \text{ bushel} = 1.25 \text{ cubic feet}$$

Dry Measure

$$2 \text{ pints} = 1 \text{ quart}$$
$$8 \text{ quarts} = 1 \text{ peck}$$
$$4 \text{ pecks} = 1 \text{ bushel}$$
$$36 \text{ bushels} = 1 \text{ chaldron}$$

Apothecaries' Weight

$$
\begin{array}{rcl}
20 \text{ grains} & = & 1 \text{ scruple} \\
3 \text{ scruples} & = & 1 \text{ dram} \\
8 \text{ drams} & = & 1 \text{ ounce} \\
12 \text{ ounces} & = & 1 \text{ pound} \\
27^{11}/_{32} \text{ grains} & = & 1 \text{ dram} \\
16 \text{ drams} & = & 1 \text{ ounce} \\
16 \text{ ounces} & = & 1 \text{ pound} \\
2{,}000 \text{ pounds} & = & 1 \text{ ton (short)} \\
2{,}240 \text{ pounds} & = & 1 \text{ ton (long)}
\end{array}
$$

Conversion Factors

Degree C = 5/9 (Degree F − 32).
Degree F = 9/5 (Degree C + 32).
Degrees absolute (Kelvin) = Degrees centigrade + 273.16.
Degrees absolute (Rankine) = Degrees fahrenheit + 459.69.

Multiply	By	To Obtain
Diameter circle	3.1416	Circumference circle
Diameter circle	0.8862	Side of equal square
Diameter circle squared	0.7854	Area of circle
Diameter sphere squared	3.1416	Area of sphere
Diameter sphere cubed	0.5236	Volume of sphere
U.S. Gallons	0.8327	Imperial gallons (British)
U.S. Gallons	0.1337	Cubic feet
U.S. Gallons	8.330	Pounds of water (20° C)
Cubic feet	62.427	Pounds of water (4° C)
Feet of water (4° C)	0.4335	Pounds per square inch
Inch of mercury (0° C)	0.4912	Pounds per square inch
Knots	1.1516	Miles per hour

Figuring Grain Storage Capacity

1 bu. ear corn = 70 lbs. 2.5 cu. ft. (15.5% moisture)
1 bu. shelled corn = 56 lbs. 1.25 cu. ft. (15.5% moisture)
1 cu. ft. = 1/2.50 = .4 bu. of ear corn
1 cu. ft. = 1/1.25 = .8 bu. of shelled corn; Bu. × 1.25 ft.³, Ft.³ × .8 bu.
Ft.³ = Bu. × 1.25
Bu. = Ft.³ × .8
Rectangular or square cribs or bins
    cu. ft. = width × height × length (W × H × L)
Round cribs, bins or silos (= 3.1416)
    Volume = R^2H = D^2H/r
    cu. ft. = × radius × radius × height = (R × R × H)
    or × diameter × diameter × height = (D × D × H)

$$\frac{}{4}$$

    or $\dfrac{3.1416 \times D \times D \times H}{4}$ = .785 × D × D × H

Figuring Grain Storage Capacity (Continued)

Examples:

1. Crib - ear corn - 6' wide by 12' high by 40' long
   a. 6 × 12 × 40 = 2880 cu. ft. × 4 bu./cu. ft. = 1152 bu.
   b. 6 × 12 × 1 = 72 cu. ft. × .4   28.8 bu./ft. of length × 40' = 1152 bu.

2. Round crib - ear corn - 14' diameter by 14' high
   a. .785 × 14' × 14' × 14' × .8 = 1722 bushel
   b. .785 × 14' × 1 × .4   6.15 bu./ft. × 14 = 861 bushel

3. Round Bin or Silo - shell corn - 14' diameter by 14' high
   a. .785 × 14' × 14' × 14' × .8 = 1722 bushel
   b. .785 × 14' × 14' × 14' × 1 × .8 - 123 bu./ft. × 14' = 1722 bushel

Metric Equivalents

Metric Weight

1 grain	=	.065 gram
1 apothecaries' scruple	=	1.296 grams
1 avoirdupois ounce	=	28.350 grams
1 troy ounce	=	31.103 grams
1 avoirdupois pound	=	.454 kilogram
1 troy pound	=	.373 kilogram
1 gram	=	15.432 grains
1 gram	=	.772 apothecaries' scruple
1 gram	=	.035 avoirdupois ounce
1 gram	=	.032 troy ounce
1 kilogram	=	2.205 avoirdupois pounds
1 kilogram	=	2.679 troy pounds

Capacity

1 U.S. fluid ounce	=	29.573 ml
1 U.S. fluid quart	=	.946 liter
1 U.S. fluid ounce	=	29,573 milliliters
1 U.S. liquid quart	=	.964 liter
1 U.S. dry quart	=	1.101 liters
1 U.S. gallon	=	3.785 liters
1 U.S. bushel	=	.3524 hectoliters
1 cubic inch	=	16.4 cubic centimeters
1 liter	=	1,000 milliliters or 1,000 cubic centimeters
1 cubic foot water	=	7.48 gallons or 62 1/2 pounds
231 cubic inches	=	1 gallon
1 millimeter	=	.034 U.S. fluid ounce
1 liter	=	1.057 U.S. liquid quarts
1 liter	=	.908 U.S. dry quart
1 liter	=	.264 U.S. gallon
1 hectoliter	=	2.838 U.S. bushels
1 cubic centimeter	=	.061 cubic inch

162

Miscellaneous Equivalents

9 in. equals 1 span

6 ft. equals 1 fathom

6,080 ft. equals 1 nautical mile

1 board ft. equals 144 cu. in.

1 cylindrical ft. contains 5⅞ gals.

1 cu. ft. equals .8 bushel

12 dozen (doz.) equals 1 gross (gr.)

1 gal. water weighs about 8⅓ lbs.

1 gal. milk weighs about 8.6 lbs.

1 gal. cream weighs about 8.4 lbs.

46½ qts. of milk weighs 100 lbs.

1 cu. ft. water weighs 62½ lbs., contains 7½ gals.

1 gal. kerosene weighs about 6½ lbs.

1 bbl. cement contains 3.8 cu. ft.

1 bbl. oil contains 42 gals.

1 standard bale cotton weighs 480 lbs.

1 keg of nails weighs 100 lbs.

4 in. equals 1 hand in measuring horses

ADDRESS OF BASIC MANUFACTURERS

1. Abbott Laboratories
   Abbott Park - 1400 Sheridan Road
   N. Chicago, IL 60064

2. Agway
   P.O. Box 4933
   Syracuse, NY 13221

3. American Cyanamid Company
   One Cyanamid Plaza
   Wayne, NJ 07470

4. Applied Biochemists
   Box 255
   Mequion, WI 53092

5. AVITROL Corp.
   320 S. Boston
   Ste. 528
   Tulsa, OK 74103

6. BASF
   Postfach 220
   6703 Limburgerhof
   West Germany (FR)

7. BASF-Wyandotte Corp.
   Ag Chemical Div.
   100 Cherry Hill Road
   Parsippany, NJ 07054

8. Bayer A. G.
   509 Leverkusen
   Bayerwerk,
   Beratung Pflanzenschutz
   West Germany (FR)

9. Bayvet Division-Miles Labs
   P.O. Box 390
   Shawnee, KS 66201

10. Bell Labs
    3699 Kinsman Boulevard
    Madison, WI 53704

11. Biochem Products
    P.O. Box 264
    Montchanin, DE 19710

12. Buckman Labs
    1256 N. McLean Blvd.
    Memphis, TN 38108

13. C.P. Chemicals, Inc.
    P.O. Box 158
    Sewaren, NJ 07077

14. Celamerck GmbH
    6507 Ingelheim
    P.O. Box 202
    West Germany (FR)

15. Chemie Linz AG
    Postfach 269
    A-4020 Linz, Austria

16. Cheminova
    P.O. Box 9
    DK-7620 Lemvig,
    Denmark

17. Chemolimpex
    Plant Protection Dept.
    P.O. Box 121
    H-1805 Budapest, Hungary

18. Chempar
    660 Madison Avenue
    New York, NY 10021

19. Chevron Chemical Company
    P.O. Box 3744
    San Francisco, CA 94119

20. CIBA-Geigy AG
    CH 4000
    Basel 7 Switzerland

21. CIBA-Geigy Corporation
    Agricultural Chemicals
    P.O. Box 18300
    Greensboro, NC 27419

22. W.A. Cleary Company
    P.O. Box 10
    Somerset, NJ 08873

164

23. Cooper McDougall &
    Robertson, Ltd.
    Berkhamsted, Herts.
    England

24. d-Con Company
    90 Park Avenue
    New York, NY 10016

25. Degesch America, Inc.
    P.O. Box 116
    Weyers Cave, VA 24486

26. Degesch GmbH
    P.O. Box 610207
    D-6000 Frankfurt 61
    West Germany (FR)

27. Dow Chemical Co.
    P.O. Box 1706
    Midland, MI 48640

28. Dr. R. Maag Ltd.
    Chemical Works
    CH-8157
    Dielsdorf, Switzerland

29. Duphar
    P.O. Box 7006
    Apollolaan 151
    Amsterdam, Holland

30. DuPont Company
    Biochemicals Division
    1007 Market Street
    Wilmington, DE 19898

31. EM Laboratories
    5 Skyline Drive
    Hawthorne, NY 10532

32. Elanco Products Company
    Division of Eli Lilly Company
    P.O. Box 1750
    Indianapolis, IN 46206

33. FBC Ltd.
    Hauxton, Cambridge
    CB2 5HU
    England

34. FMC Corporation
    Ag Chemical Div.
    2000 Market Street
    Philadelphia, PA 19103

35. F. W. Berk & Co., Ltd.
    8 Baker Street
    London W1, England

36. Fairfield American
    238 Wilson Ave.
    Newark, NJ 07105

37. Farmoplant S.P.A.
    Piazza Della Republic 6
    20124 Milano, Italy

38. Great Lakes Chem. Corp.
    P.O. Box 2200
    W. Lafayette, IN 47906

39. Griffin Corporation
    P.O. Box 1847
    Valdosta, GA 31601

40. Gustafson, Inc.
    17400 Dallas North Parkway
    Dallas, TX 75252

41. Guth Corporation
    332 S. Center Street
    Hillside, IL 60162

42. Hercon
    1107 Broadway
    New York, NY 10010

43. Hodogaya Chemical
    1-4-2 Toranomon-1-Chome
    Minatoku-Tokyo 105
    Japan

44. Hoechst-Agrochem. Div.
    Postfach 900320
    6230 Frankfurt (m) 80
    West Germany (FR)

45. Hoechst-Roussel Agri.
       Vet Co.
    Rte. 202-206 North
    Somerville, NJ 08876

46. Hokko Chemical Industries
Mitsui Building 2
4-2 Nihonbashi Hongoku-Cho
Tokyo 103 Japan

47. Hopkins Ag Chemical Co.
P.O. Box 7532
Madison, WI 53707

48. ICI Americas, Inc.
Wilmington, DE 19897

49. ICI Ltd.
Plant Protection Div.
Fernhurst, Hasslemere
Surrey, England GU27 3JE

50. Ihara Chemical Co.
1-4-26 Ikenohata-1-Chome
Taitoku Tokyo, 110 Japan

51. Janssen Pharmaceutical
P.O. Box 344
Washington Crossing, NJ 08560

52. Janssen Pharmaceutica N.V.
Agricultural Div.
B-2340 Beerse, Belgium

53. Kaken Chemical Co., Ltd.
No. 28-8, 2 Chome
Honkomagome, Bunkyo-Ku
Tokyo 113 Japan

54. Kalo Labs
4550 W. 109th St.
Suite 222
Overland Park, KS 66211

55. Kanesho Company, Ltd.
Rm. 333, Marunouchi Bldg.
Maunouchi, Chiyoda-ku
Tokyo, Japan

56. Keno Gard AB
P.O. Box 11033
S-10061
Stockholm, Sweden

57. Kumiai Chemical Industries
4-26 Ikenohata
Tokyo 110 Japan

58. Kureha Chemical Ind. Co.
1-9-11 Nihonbashi,
Horidome-cho
Chuo-ku,
Tokyo 103 Japan

59. Maag Agrochemicals
P.O. Box 6430
Vero Beach, FL 32961

60. Makthestim - Agan
P.O. Box 60
Beer-Sheva,
Israel

61. Mallinckrodt, Inc.
Specialty Chemical Div.
P.O. Box 5439
St. Louis, MO 63147

62. May & Baker Ltd.
37-39 Manor Road
Romford, Essex
2MI 2TL England

63. McLaughlin Gromley King
8810 - 10th Ave., North
Minneapolis, MN 55427

64. Meiji Seika Company
4-16 Kyobashi 2-Chome
Chuo-ku Tokyo, 104 Japan

65. Merck & Company
P.O. Box M
Rahway, NJ 07065

66. E. Merck A. G.
61 D Armstadt
Franfurter, Strasse 250
West Germany (FR)

67. Miller Chemical & Fert. Corp.
Box 333
Hanover, PA 17331

68. Minerals Res. & Devel. Corp.
4 Woodlawn Green
Suite 232
Charlotte, NC 28210

69. Mitsubishi Petrochemical Co. Ltd.
    5-2 Maraniuchi 2-Chome
    Chiyoda-ku Tokyo, Japan

70. Mitsui Agricultural Chemicals
    2-5 Kasumigaseki 3-Chome
    Chiyodaka,
    Tokyo, Japan

71. Mitsui Toatsu Chemicals
    P.O. Box 83
    Kasumigasehi Building
    Tokyo 100 Japan

72. Mobay Chemical Co.
    P.O. Box 4913
    Kansas City, MO 64120

73. Monsanto Chemical Company
    800 N. Lindburgh Blvd.
    St. Louis, MO 63166

74. Motomco Ltd.
    P.O. Box 6072
    Clearwater, FL 33518

75. Murphy Chemical Ltd.
    Latchmore Court, Brand St.
    Hitchin, Herts SG5 1H2
    England

76. Nalco Chemical Company
    2901 Butterfield Rd.
    Oak Brook, IL 60521

77. Nihon Nohyaku Company, Ltd.
    2-5 Nihonbashi 1-Chome
    Chuo-ku Tokyo 103 Japan

78. Nihon Takushu Noyaku Seizo
    Honcho Bldg.
    2-4 Nihonbashi
    Chuo-ku
    Tokyo 103 Japan

79. Nippon Kayaku Co.
    Fujimi Bldg. 11-2 Fujimacho
    1 Chome, Chiyoda-ku
    Tokyo 102 Japan

80. Nippon Soda Co., Ltd.
    Ag Chemical Div.
    P.O. Box 1173
    Tokyo, Japan

81. Nissan Chemical Ind., Ltd.
    Kowa-Hitotsubashi Bldg.
    7-1, 3-Chome, Kanda
    Nishiki-cho
    Tokyo, Japan

82. Nor-Am Ag. Products, Inc.
    3509 Silverside Rd.
    Wilmington, DE 19803

83. Nordox A/S
    Ostenjovn 13
    Oslo, 6 Norway

84. Occidental Chemical Co.
    P.O. Box 1185
    Houston, TX 77001

85. Otsuka Chemical Co.
    10 Bungo-Machi
    Higashi-ku 540
    Osaka, Japan

86. PPG Industries
    One Gateway Center
    Pittsburgh, PA 15222

87. Penick Corp.
    1050 Wall Street West
    Lyndhurst, NJ 07071

88. Pennwalt Corporation
    Ag Chemical Div.
    3 Parkway
    Philadelphia, PA 19102

89. Pepro
    B.P. 139
    69 Lyon R.P.
    France

90. Pestcon Systems
    P.O. Box 469
    Alhambra, CA 91802

91. Pfizer Inc.
    235 E. 42nd St.
    New York, NY 10017

92. Phelps Dodge Refining Corp.
    300 Park Avenue
    New York, NY 10022

93. Phillips Petroleum
    Bartlesville, OK 74004

94. Prentiss Drug &
    Chemical Co., Inc.
    21 Vernon St.
    CB 2000
    Floral Park, NY 11001

95. Ralston Purina Company
    Checkerboard Square
    St. Louis, MO 63188

96. Rentokil Laboratories
    Felcourt, East Grinstead
    Sussex, England

97. Rhone-Poulenc
    P.O. Box 125
    Monmouth Junction, NJ 08852

98. Rhone-Poulenc
    B.P. 9163
    69263 Lyon Cedex I
    France

99. Rohm & Haas Company
    Independence Mall West
    Philadelphia, PA 19105

100. Roussel UCLAF
    163, Ave. Gambetta
    75020 Paris, France

101. Sandoz, Ltd.
    Agrochemical Department
    Basel, Switzerland CH-4002

102. Sankyo Co., Ltd.
    No. 7-12, Ginza 2-Chome
    Chuo-ku
    Tokyo 104 Japan

103. SARIAF
    20124 Milano
    Italy

104. Schering AG
    Postfach 650311
    Berlin 65, West Germany (FR)

105. SDS Biotech Corp.
    P.O. Box 348
    Painesville, OH 44077

106. SDS Biotech KK
    No. 2 Higashi Shinbashi Bldg.
    12-7 Higashi Shinbashi 2-Chome
    Minato-ku
    Tokyo 105 Japan

107. Shell Chemical Company
    P.O. Box 3871
    Houston, TX 77002

108. Shell International
    Chemical Company
    Agrochemical Division
    Shell Centre
    London SE1 7PG
    England

109. Shionogi and Co., Ltd.
    12-Sanchome, Doshomachi
    Osaka, 541 Japan

110. Stauffer Chemical Div.
    Agricultural Chemical Div.
    Westport, CT 06880

111. Sumitomo Chemical America
    345 Park Avenue
    New York, NY 10022

112. Sumitomo Chemical Co., Ltd.
    15 5-Chome Kitahama
    Higashi-ku, Osaka
    Japan

113. Takeda Chemical Industries
    12-10 Nihonbashi 2-Chome
    Chuo-ku
    Tokyo 113 Japan

114. Tennessee Chemical Co.
     3475 Lenox Rd., NE
     Suite 670
     Atlanta, GA 30326

115. 3-M Company
     Agricultural Chemical
     Products Bldg. 223
     3-M Center
     St. Paul, MN 55144

116. Union Carbide Ag Prod.
     P.O. Box 12014
     Research Triangle Park, NC 27709

117. Uniroyal Chemical
     Agricultural Chemical Division
     320 Spencer St.
     Naugatuck, CT 06770

118. Velsicol Chemical Corporation
     341 E. Ohio Street
     Chicago, IL 60611

119. Vertac Chemical Corporation
     5100 Poplar Avenue
     Suite 2414
     Memphis, TN 38137

120. Vineland Chemical Company
     West Wheat Road
     Vineland, NJ 08360

121. Wacker-Chemie GmbH
     Prinzregentenstr. 22,
     800 Munchen 22,
     West Germany (FR)

122. Zoecon Corporation
     975 California Avenue
     Palo Alto, CA 94304

NOTES